SACRED GEOMETRY

· LANGUAGE OF THE ANGELS ·

"Heath illuminates the cosmological ideas informing the teachings of John Bennett and George Gurdjieff. His book is a masterful insight into the origins of ancient wisdom, presented with impeccable analyses of megalithic and historical monuments. It gives a theory of angelic or higher intelligence based on astronomy, music, and number—reflecting the 'deep thought' operating at the start of human consciousness."

ANTHONY BLAKE, AUTHOR OF
A GYMNASIUM OF BELIEFS IN HIGHER INTELLIGENCE
AND *THE INTELLIGENT ENNEAGRAM*

"In *Sacred Geometry,* Richard Heath proposes, based on his deep and comprehensive analysis of primary sacred sites, that numbers and geometries are the primary medium for divine creation. He *proves* these sites were built based on specific numbers and geometrical patterns that correspond with the musical octaves. Heath concludes angels inspired these sites to advance evolution on Earth, a very radical idea. If this is true, then angelic science guided the numeric order built into Earth's design and it's unfolding in time. I think he's right. *Sacred Geometry* is a deep and sustained meditation with the mind of Gaia—brilliant!"

BARBARA HAND CLOW, AUTHOR OF
THE MIND CHRONICLES AND
ALCHEMY OF NINE DIMENSIONS

SACRED GEOMETRY

· LANGUAGE OF THE ANGELS ·

RICHARD HEATH

Inner Traditions
Rochester, Vermont

Inner Traditions
One Park Street
Rochester, Vermont 05767
www.InnerTraditions.com

Cataloging-in-Publication Data for this title is available from the Library of Congress

ISBN 978-1-64411-118-5 (print)
ISBN 978-1-64411-119-2 (ebook)

Printed and bound in China by Reliance Printing Co., Ltd.

10 9 8 7 6 5 4 3 2 1

Text design and layout by Debbie Glogover
This book was typeset in Garamond Premier Pro with Majesty, ITC Galliard Std, Posterama Text, and Gill Sans MT Pro used as display fonts

Because hyperlinks do not always remain viable, we are no longer including URLs in our resources, notes, or bibliographic entries. Instead, we are providing the name of the website where this information may be found.

To send correspondence to the author of this book, mail a first-class letter to the author c/o Inner Traditions • Bear & Company, One Park Street, Rochester, VT 05767, and we will forward the communication, or contact the author directly at **sacrednumber@gmail .com** or through **https://sacred.numbersciences.org**.

Contents

PART TWO

THE COSMIC INDIVIDUALITY

Preface

SACRED BUILDING OVER THE LAST five to seven thousand years often referenced a small set of highly specific designs, and these appear to be cosmic prototypes for the creation of life on Earth. This suggests that numbers and geometries are the primary medium for the Universal Will,* just as physical laws are now known to determine the functional aspects of the universe. Laws enabled there to be stars and planets, but it was special "sacred" numbers that created the cosmic conditions for life.

The creation of the universe is complementary to its evolving self-consciousness. The present moment contains situations that can be understood through a limited number of terms whose interrelationships lead to a new capacity to act from understanding. The traditional symbol of this structure-building capacity was the wheel with a number of spokes, resembling the tone circle of an octave. This arose through a number of spiritual initiatives around 600 BCE, marking the coming ascendancy of the Cosmic Individuality, a name given to this evolving self-consciousness of the universe, a savior from its otherwise endless material expansion and ultimate doom. The Purpose of the universe is only to be found and fulfilled in this Cosmic Individuality, which is the reflex image and likeness of God, and is the beginning of a new type of creation on the Earth and other similar planets endowed with a large moon.

◆ ◆ ◆

---

*Crucial to notions that the universe was in some way created is that a creator willed it. Cosmologies of the traditional kind often then go further, to see the structure of the universe as reflecting that Universal Will as to how it should be, perhaps even a work in progress, and not a finished product.

The story of this book would not have been possible without the work of Alexander Thom (1894–1985), who surveyed and interpreted the megalithic sites found in Britain and Brittany, and also the work of John Michell (1933–2009), who realized the importance of metrology and geometry to the interpretation of ancient buildings and the mysteries these represent. By 1991, my brother Robin Heath had rediscovered the all-important Lunation Triangle geometry from which appears to have flowed the form of numeracy responsible for the megalithic civilization of the Atlantic coast of Europe. By 2000, John Neal (a colleague of John Michell) had reconstituted the likely form of ancient metrology, essential for understanding the original design concepts behind ancient buildings, and he had also recovered the accurate model for the Earth's size and shape known in antiquity. Howard Crowhurst introduced me to the world-class megalithic complex around Carnac in Brittany, providing tours in 2004 and 2007, and copies of rare *Études et Travaux* magazines (from the late 1970s and 80s), from a past member of the "AAK" study group.

The overall context for the book only became clear upon reading the fourth volume of John G. Bennett's (1897–1974) *The Dramatic Universe: History.* Bennett's proactive approach to Gurdjieff's work on ideas facilitated my own contributions in chapter 9. My friendship with Anthony Blake, Bennett's pupil, helped refine my work on ideas and on significant numbers that enable the world to be understood in both ancient and developmental ways. In similar fashion, I could not have seen the significance of celestial harmony, first noted in my *Matrix of Creation,* without years of distance learning with Pythagorean musicologist Ernest G. McClain (1918–2014) regarding the harmonic number symbolism of the Bible and Plato, outlined in my 2018 *Harmonic Origins of the World.*

As usual, this book is greatly enhanced by the professionalism of the publishing team at Inner Traditions. Any remaining errors might be due to the scale of the story itself for any individual attempting to tell it. I am indebted to each of those mentioned for sharing knowledge, research, and images with me so that I may do the same with the publication of this book. Though not otherwise noted, all images have been used with permission, and I am grateful to each researcher and his or her family or institution for the trust placed in me to use the images responsibly.

I hope some readers will calculate for themselves and so understand the power of late Stone Age numeracy, which profoundly influenced the later civilizations, leading to our own.

Angelic Transmission through Sacred Number

THE IDEA OF ANGELS has recently broken free of an orthodox religious context, and angels are now seen as higher beings who care for human lives without any ecclesiastical sanction. Such beings are different from mere spirits, being responsible for an overall evolution of life on Earth: its entire scope of past, present, and future. This planetary role of angels is related to the emerging destiny of human beings, whose minds hardly comprehend the scope of the enterprise they are within. From an angelic perspective, the human race is worth helping in order to progress with their own work.

The angelic mind is founded on a deep understanding of numbers and the patterns they produce, patterns that belong to a *universal framework science* that defines what is possible and impossible within planetary worlds and their specific situations. These framework conditions caused this universe, and they have governed its development according to the Universal Will. Humans now have added their minds to the phenomenon of life on Earth, and these minds must have resulted from the framework conditions once life had evolved human beings.

That minds should arise on Earth is then, probably, no accident. Rather, they are essential to the purpose of the human essence class itself. Unlike the working of angels, who facilitate a living planet's creation, living minds enable the universe to know itself. We are microcosms, or small resemblances of a greater whole, that can know nothing without our minds. The universe—macrocosm or greater whole—must grow minds to become self-aware. The minds of angels,

1

on the other hand, differ from ours in their furthering of Earth's evolution: angels are part of the Universal Will to create the universe. In contrast, human minds are part of the developing self-awareness of the universe and a stepping-stone toward human transformation into the Cosmic Individuality.

Though an exact science has emerged through human civilization, this is significantly different from the universal science of angels. For a start, the mind of an angel differs from that of a human; it is inherently top-down, thinking as it does about harmony within the universal framework conditions. Human thinking is instead bottom-up, being driven by contingency and with no direct insight into the cosmic order apart from what can be deduced through the human mind and its sensorium.

For angels to encourage the development of human minds, they had to reach the human sensorium with a complex but intelligible message, one found within a suitable sensory phenomenon. Fortunately, the angelic world had already created, out of necessity, a suitable phenomenon for communicating with early human minds that involved number patterns and language: the planetary world had already been organized according to angelic number science. Therefore, around six thousand years before the present, *Homo sapiens sapiens* could become aware of this phenomenon through the astronomical counting of planetary time periods. A prehistoric human science then emerged that was based on angelic science and that would be influential in developing the human mind and civilization.

The planetary world had been made intelligible through the creation of the Moon and its resonances with the other planets. This had been necessary for life to arise on Earth, but these resonances could now be recognized from the Earth by the simple counting of days and could be seen as being based on the properties of numbers. Astronomy was therefore able to transmit angelic number science—in particular numeracy, rationality, and reasoning—to late Stone Age humans so as to develop their minds. These humans then also used numbers to form the megalithic measures for designing astronomical observatories, and they eventually evolved a highly integrated system of measures and geometrical forms with which to achieve what we now achieve through arithmetic and trigonometry. This ancient science of measures continued alongside the development of later arithmetical and geometrical methods, but over time became increasingly restricted to being used in the sacred buildings constructed by its remaining initiatory groups.

After the basic astronomy of the solar and lunar years was established, musical ratios to the lunar year were also noticed between the lunar year and the planetary bodies. The shape of the Earth itself was found to conform to three rational approximations to the number *pi* (represented as π), again without any need for humanity's later scientific methods. In other words, the surrounding circumstances of the Earth already presented a multifaceted cosmogram, allowing angelic science to be quantified on Earth by a late Stone Age culture using megaliths, reflecting what angelic science had achieved in the process of making Earth a living planet. Coming into rapport with the past work of angels, the early mind experienced itself, and this echoed forth into many different religious cultures and secret guilds, only then to be misunderstood as having involved a supernatural rather than a mental contact, the latter mediated by the astronomical work.

This arising of early number sciences has placed an enigma within our prehistory. We cannot explain the sudden arrival of numeracy by 3000 BCE in the ancient Near East and perhaps elsewhere. Religions also sprang up based on myths, containing an underlying symbolism of sacred numbers derived from angelic science. Often now denigrated as being based on a kind of animism, superstition, or similar conceit, such religions separate modern science from its origins in the megalithic. Science has developed especially deaf ears toward past religious controls over the mind of man. It also cannot believe that megalithic astronomy was fully an exact science, though methodologically different from modern science. For science, the universe is the result of its initial conditions. Such an "accidental universe" is without any intention as to how it should develop and therefore has no need of angels or a spiritual science based on the properties of number.

In short, ancient number science is eminently deniable but, in denying its existence, science has failed to see the numerical order built into the design of the Earth and its harmonic time environment. Furthermore, it has assumed that the human mind emerged purely from the natural existence of primates living in the biosphere. Science and scholarship need to understand essential facts about the Earth's environment, that it was clearly created to be simple while sophisticated and that this made the maximum impression on the megalithic astronomers who effectively marked the end of prehistory and the beginning of civilized history.

There is little danger of a return to religious thinking as originally conceived. The human mind, shaped by modern science, has reached the natural limits of its own paradigm, in which there can be no causes other than material forces. Isolation from the spiritual world, which created the planet and its framework for life to emerge, prevents the mind, as that science, from going any further because the next stage essentially requires the discovery that our planetary world is an artifact of higher intelligence. Exploitation of the Earth's resources then appears stupid.

The prehistoric debate in so many books, which proposes every kind of alternative history to explain the enigma of mankind's sudden possession of civilization, is a symptom of the fixed paradigm of science. The prehistoric number sciences, first found in the megalithic period, are required to explain this enigma, since they were the means through which humans developed minds that were capable of forming an early exact science, minds that then went on to form our present-day science.

PART ONE

THE UNIVERSAL WILL

The English word *universe* is made up of *uni,* meaning "one" and *verse,* meaning "turn." Its traditional mythos is that of a creator whose creation returns to its source as something that has been changed. This is somewhat different from a merely creative god, so that, in approaching this idea, the Indians invented three gods—ones unlike the Christian Trinity—called Brahma, who creates, Vishnu, who maintains, and Shiva, who destroys the creation. This enables one to trace the creation, maintenance, and destruction of the manifested universe as a circular journey, a natural form of narrative that these days is called a ring composition. But the god who would initiate such a universe does so with a purpose that may be called the Universal Will for this universe. As with any purpose in life, success comes through the originating will being present not as a local divinity but as a principle reflected in all that happens within the stages and processes of creation, maintenance, and dissolution on many necessary levels.

It seems that the Universal Will includes the possibilities for life on Earth that have made the plants and animals and then the humans and their ability to frame all these possibilities in their minds. All living things depend on each other more than do rocks and the inanimate world in general, and life can be seen as an infinite set of reflex experiences, so that the interdependencies of living things appear to be a principle on which the universe with all its interactions was established by God. This cybernetic principle of feedback was very popular in the decades after the Second World War. Now seemingly forgotten, it has been assimilated by later ideas such as chaos and catastrophe theory, which are more formulaic and less profound. The ultimate manifestation of cybernetics was in the anthropic principle, which proposed that the universe was finely tuned to produce stars, planets, and life through its laws and chemical composition. Some went further in proposing a stronger version of it, that the universe was organized not only to create life but also to create the self-awareness of thinking beings, who could then complete the universe through being witnesses to what has been and is being created, through the human experience. That is, God created the universe to become self-aware.

The relationship between "God and Man" was a late formation developed by the monotheistic religions. It led, through denial of an intermediary *spirit* world that affects phenomena, to modern science, technological and industrial progress, and then consumerism. But postreligious humans could no longer think of themselves as kind productions of God, or of many gods, but rather as beings evolved by a natural selection that surely the divine did not invent. Yet, of course, interdependence is still everywhere present in the universe and strongly within life, and so God has been reduced to being an agony aunt for the religious or has joined cybernetics as forgotten in its deeper significations* due to a failure to see what is the reality of the intermediary higher world of eternity. It is this that concerns us here since in prehistory an extraordinary orderliness within the planetary world allowed late Stone Age people to achieve miraculous knowledge, using only common sense plus the properties of integer numbers and easily constructed geometrical forms. This was a purely geocentric phenomenon, only seen from the surface of the Earth. The only possible reason for this geocentricity was that the planetary world and form of the Earth was a construction by creative minds we would call angels.

Angels have the role of bringing about the manifest universe through their direct imagination of ratios and geometries. Once the megalith builders, in their astronomical journey, started using (a) alignments to the Sun and Moon, (b) the counting of days in longer cycles, and (c) the comparison of results within geometrical forms, humans developed minds similar to but different from those of angels.

This initiation of the human mind was through sympathetic contact with the "sky gods," leaving myths that explained how humans were initiated by the gods or by a progenitor of knowledge. The initiate became the messenger of God's angelic achievement, and from this contact with the Universal Will, a unified system of geometrical models and measures arose and survived in many regions, up until modern times, because many sacred buildings have survived. These ziggurats, pyramids, domes, and so on, modeled the Earth and preserved information in the exact lengths and measures used to fashion them. In the Bible's book of Genesis, men built a Tower of Babel, a ziggurat that would reach unto the astronomical heavens, so that Jehovah destroys it, causing the one language (of number) that men spoke to become many scattered languages,

---

*The widely quoted statement by Friedrich Nietzsche, "God is dead," was figurative yet has become emblematic of modernism's rejection of the need for God to explain the world. I am sure God would agree with Mark Twain that "the report of my death was an exaggeration," though of course God would not be a being who could die.

so that they should not understand as God or an angel does. This one language of angels, of geometry and metrology, that built sacred structures to represent the Earth as a sphere, the Moon, the planets, their harmony, and all the knowledge and understanding, developed from the angelic mind. In myth, angels are sometimes blamed for giving humans their minds.

Therefore, part 1 is an exploration of this tradition, which passed to groups that could still build using the old techniques, but whose knowledge became secret and in some cases suppressed, possessed exclusively by specialists who often served only kings and priests in making high-status monuments, meaningful in religious terms but for unknown reasons.

A building based on objective truths has a subconscious effect on the culture that possesses it because that building belongs to the higher worlds and yet is among us. The human subconscious is like the mythical Atlantis and its mythos of lost knowledge. It has sunk beneath the waves of an ocean of unconsciousness, still functioning but now subconsciously. Like a sunken vessel, much of its construction is unfamiliar, yet, like a flying saucer in modern myths, the Thing (as northern folk called it) affects our dreams and, like King Arthur, awaits the return of consciousness. In our present minds, it is quite baffling how so many buildings have been made over the last five thousand years that tell the same few stories in many different byways of an ancient art derived from the work of angels.

And almost everything depends on the Moon, whose creation seems to have made the Earth the progenitor of life. According to Skeat's *Etymological Dictionary of the English Language,* the Moon's etymology derives from the Sanskrit *me* meaning "to measure, as it is a chief measurer of time," and the lunar month and year were the first subject of study, and a constant companion for late Stone Age astronomers.

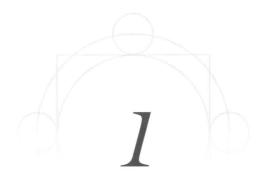

1

Ratios of the Angelic Mind

HUMAN NUMBER SCIENCE FIRST AROSE because of early astronomers who intuited a set of enigmatic geometrical structures within time and space. Structures have survived that embody these geometries, such that one cannot avoid finding that science within monuments such as the pyramids of Egypt and stone circles such as Stonehenge. Our views of these early technologists are that they had no technology since they had not the power to develop one. Official science continues to assert that the culture that built these monuments had an animistic worldview, from which there could not have emerged the sort of big picture that modern science now has of the world. Yet our science warns us of the power of confirmation bias, where *what we expect we are liable to see;* and this has led to widespread prejudice that the early astronomers were Stone Age primitives in their beliefs. Yet many of our own fictional works have a recurrent theme in which we find an alien or ancient science or magic that completely changes the modern world. Reality was just like that, which is perhaps why our subconscious or group mind also has a strange attraction to this plot about adventures through archaeological monuments.

The payload for this "alien intervention" into the mind of late Stone Age humans came from the sky in the form of a detectable set of numerical relations imposed on the Earth, the Moon, and the planets in order to create life on Earth. In detecting these relationships, early humans came into contact with an angelic science involving numbers, and hence, the megalith builders became numerate, but in a way different from today's numeracy, which involves a mathematics that writes numbers and does arithmetic, trigonometry, analytical geometry, calculus, and so on. Any contemporary person is naturally challenged to comprehend this angelic number science, yet it was simple in its founding concepts.

Hidden structures exist within the linear set of cardinal numbers (1, 2, 3, 4, 5, 6, 7, etc.), the bread and butter of modern calculations. The argument is powerful that without historical arithmetic, cultures would have stood no chance of making calculations. Fortunately, angelic science was like the goddess Athena, "born with a shout, fully clothed and armed" during the creation of the early universe as a set of methods in which calculation is done by ratios in which numbers appear as multiples (numerators) and submultiples (denominators or divisors).

On today's calculators we would simply enter a number or quantity, Q, which we can then multiply and divide. By the late Stone Age, the quantity L had arisen through counting time as a number of days, using a fixed unit of length. The difference between the modern and ancient approaches to calculation is that numbers could not be abstracted. So without written numbers, the megalithic would have to know the numbers contained within a length using some form of analysis. The three main categories of such analytical techniques are:

1. Factorization: To apprehend the numerical content of a length required experiments in which small numbers (say 5 inches) were exhaustively divided into L and, when a whole number of 5 resulted, it became known that the length could be divided by 5. In other words, the Stone Age had to use factorization of a number to build up an idea of what a number contained.

2. Rescaling: Initially, a right triangle could be used whose longest sides differed by one common unit. For example, L could be the length of the base side and, by dividing L into 8 portions, a rope one portion greater would be 9 units long. If that rope was arced until the length of 9 stood above L, then each unit on the base would have a longer unit above it that was 9/8 longer.

3. Comparison: When two cosmic time periods needed to be compared, then as with the above, the longer could stand above the shorter length to again form a right triangle. Arcing the longer back down to the baseline, the difference in lengths could be found to create a new unit that

then divides into both base and hypotenuse so as to normalize the triangular ratio as being $L = N$ (the base) and $L = N + 1$ (the hypotenuse).

THE REQUIRED ACT OF SELF-INITIATION

Megalithic monuments became a record of what was being measured, of how astronomers were interacting with those measurements, and of horizon events, to understand how the time world was structured. The starting point was to clearly understand the relationship between the solar and lunar years using the comparison method. What was up there, in the time world, would prove to be crucial. The angelic mind had constrained the visible manifestations of the Sun and Moon to happen in a relatively short periodicity of nineteen years, the period called Metonic by the Greeks. This was achieved by aiming for a 19-year period that would also be an anniversary for the number of lunar months in 19 years, namely 235 months. In table 1.1, the first row expresses the one-year, three-year, and nineteen-year periods in lunar months (LM):

TABLE 1.1
THREE SIGNIFICANT TIME PERIODS
COUNTED IN LUNAR MONTHS, DAY-INCHES, AND
MEGALITHIC YARDS (MY) PER MONTH

Units	One solar year	Three solar years	Nineteen solar years
Lunar Months	12 + **7/19 LM**	36 + **21/19 LM**	228 + **7 LM**
Day-inches	354.3" + **10 7/8"**	1063.1" + **32 5/8"***	6733 + **206.7"**
Month (MY)	12 + **7/19 MY**†	36 + **21/19 MY**	228 + **7 MY**

*This is the creation of the first megalithic yard at Carnac, visible within the Le Manio Quadrilateral (fig. 1.2, p. 14).

†This is the creation of the first English foot, also visible on stone C3 of the Gavrinis chambered cairn (fig.1.9, p. 30).

The nineteen-year coincidence with 235 lunar months has significant consequences for both single-year and triple-year counting. Over a single year (row 1), the solar year must have 7/19 (0.368) lunar months as the difference between it and the lunar year. Over 3 years though (row 2), the number of day-inches between the two years equals a megalithic yard (MY) that, divided by 12, is very close to 19/7 feet long. This enabled the megalithic yard to be used to count lunar months (row 3), causing the length 7/19 MY to be 12 inches long—our foot length—due to the cancellation of 7/19 by 19/7, the megalithic yard. And it is this length that showed that the mean solar month (one-twelfth of the solar year) was then one inch longer than the lunar month when counted in MY. This initiated many future instances of twelve-ness, since the solar year then came to be seen as 12 solar months, just as the lunar year has 12 lunar months; and these solar months appear in lengths above the base lengths of the 12 lunar months below, on the triangle's base. Furthermore, the Zodiac became a natural way to divide up the stars in a 12-fold band around the Sun's path (today called the Ecliptic).

What should have been a complex piece of astronomy was therefore collapsed into just these two steps, of day-inch counting over three years and of counting lunar months over a single year in megalithic yards. It was the 19-year constraint over lunar behavior (235 months) that made this feasible for late Stone Age people. In this, one may quote Fred Hoyle when he said (regarding his discoveries of how the cross-sectional areas of atoms change within stellar nucleosynthesis*): "A common sense interpretation of the facts suggests that a superintellect has monkeyed with [the numbers], and that there are no blind forces worth speaking about in nature. The numbers one calculates from the facts seem to me so overwhelming as to put this conclusion almost beyond question."[1]

While the angelic mind could directly see the ratios of the time world, the right-angled triangle gave the early human world access to the fundamental ratios within time. Since their storage of numbers were in physical lengths, their metrology then evolved to *replace the right triangle* (for example, with a slope ratio such as 9/8), with a comprehensive set of ratio-based measures, based upon the English foot (such as one 9/8 feet long). In this example, of multiplication by 9/8, an existing count could be recounted in 9/8 foot units and then read in English feet—English feet being a standard foot representing the number one (see next section and appendix 1). That is, metrology displaced triangular geometry once feet of different ratios offered the same ability to reproportion

---

*Without stars born of nucleosynthesis the universe would be unable to develop as it has.

lengths and similarly analyze them as to their factors or to rescale one problem into a more amenable form or size.

Geometry is defined by the relative size of its visible lengths, as in the sides of a triangle or by invisible lengths such as the radius of a circle. Early astronomy counted the days taken for a recognizable time cycle (such as three solar years) to complete using a constant unit of length to count each day,* and these units accumulated to yield an invariant number (such as 1095.75 day-inches). As in the comparison method above, different lengths could be stored as ropes to compare them within a right triangle.

Today, such measurements seem unlikely to reveal new fundamental discoveries, but in this case our science has failed to remake these megalithic discoveries exactly because of our ways of calculating (through the mathematics of celestial dynamics) and also because of our strong belief that the world came into existence *without higher intelligences*. Because of this, the property of length is rarely studied in megalithic monuments, except when documenting their size, for posterity, in meters. Monuments are even taken apart for restoration, losing the exact metrology upon which they were built.

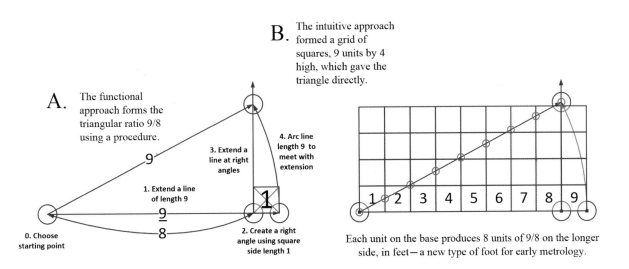

FIGURE 1.1. The ratio 9/8 seen as a triangular geometry.

---

*Length is still synonymous with time, as shown by the phrase "How long will it take?" Note also that in day-inches, large fractions, such as quarters or eighths, could be estimated as fractions of both the day and of the inch.

READING THE ANGELIC MIND

It is a sad fact that some of the most important ancient sites are underappreciated and little attended to, one such being the Le Manio Quadrilateral near Carnac. This geometry still expresses the angelic science that created the Sun-Moon-Earth system, first revealed in the 3-year cycles of the Sun, the Moon, and eclipses (fig. 1.2). Still used today, the inch was employed to count days as day-inches, and the excess of three solar years over three lunar years was found to be what we now call the megalithic yard (MY) of 3 × 10.875 day-inches = 32.625 day-inches, explored later as the Lunation Triangle (chapter 2).

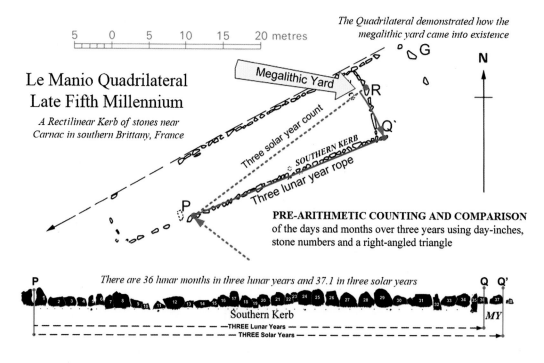

FIGURE 1.2. The use of day-inch counting and triangular comparison of time periods at the Le Manio Quadrilateral, circa 4000 BCE: the plan of the quadrilateral showing the Lunation Triangle and counted year lengths (*above*), and the silhouette of its Southern Kerb from the solstice "sun gate" and stones as lunar months in three years (*below*).

This incorporates the relevant part of Thom and Thom,
Megalithic Remains in Britain and Brittany, *1978, fig. 9.11.*
Also featured in Heath (2018), 16.

As stated above, the English foot of 12 inches was first created by using this MY, instead of day-inches, to count lunar months. When a month-MY count was made over a single year, the excess of 10.875 days became the foot of 12 inches, and the triangle was then also revealed in its normalized form in feet* as the ratio of 32.585 feet to 33.585 feet, so that, astronomically, the true value of the MY is 32.585 inches or 2.7154 feet. This new MY has been called the astronomical megalithic yard (AMY).[†]

The quadrilateral in figure 1.2 could mark the first time this day-inch count over three years was ever conducted. The MY it generates (32.625 day-inches) is probably the progenitor of the megalithic yards found by Alexander Thom in his British surveys (1938–1978) and then his Carnac surveys (1972–1973), but, at Le Manio, Thom had no knowledge of day-inch counting and hence of the characteristic unit seen, since 2010,[2] as emerging from the technique of day-inch counting (fig. 1.3). This megalithic yard naturally became symbolic of the Moon itself and its Lunar Month in particular and, by the third millennium in Carnac, the value of the foot can be seen in one of the end stones of the Gavrinis chambered cairn 4 kilometers to the east (c. 3500 BCE, see fig. 1.9, p. 30).

When the MY generated at Le Manio was used to count lunar months over a single year, *instead* of counting in day-inches, the lunar year of 12 MY led to a differential length, between the years, of the English foot. Ancient metrology went on to use this foot (only recently called English) to represent the number 1, within a cunning system for calculating using ratios without either the arithmetic of later millennia or the triangular geometry of the megalithic period. Representing ratios as feet required that such a unit, larger

---

*Normalization of a right triangle reduces the measurements of a structure to its simplest form. For example, a {3 4 5} triangle could be {12 16 20}, but when divided by 4, it is what we call it, a {3 4 5} triangle: the two longest sides differ by one whole unit. All right triangles can be normalized by dividing their two longest lengths by the difference in their length, which is then by this reduced to one whole unit. In the example, the longest lengths, {16 20}, were different by 4. Normalization can work on non-integer triangles to always give the simplest form of them, of $N:N + 1$ (called a superparticular ratio, e.g. {4 5}). In the Lunation Triangle, the normal form between the lunar year and the solar year, when normalized over nineteen years, is {32.585 33.585} and $N = 32.585$ or, over three years, 32.625 day inches—the megalithic yard first realized at Le Manio.
†Astronomical megalithic yard (32.585 inches) is a term introduced by Robin Heath to differentiate it from other metrological megalithic yards. The story of the early megalithic yard is given in appendix 1.

than the inch, be used and, when megalithic yards counted whole months, the required foot was the excess generated between a single solar year and a single lunar year.

The right-triangular geometry, for comparing geocentric time cycles, naturally led to a system of metrology where the implicit properties of such triangles, to quantify relative proportions, was displaced by a nongeometrical system of proportionate measures. But geometrical right triangles again provided the means to create those new types of foot, for any given ratio relative to the English foot. For example, a base equal to 8 feet and a longest side of 9 feet shows that above every foot on the base lies a stretched foot 9/8 times longer* (as per fig. 1.1 on p. 13).

The astronomers-turned-metrologists figured out that these stretched or compressed feet, once available, superseded their use of triangles by creating a range of foot measures that could be recorded using measuring sticks or rods.

FIGURE 1.3. Le Manio Quadrilateral, its "sun gate" (aligned to both summer and winter solstice sunrises), and the key lengths measured, revealing that a Lunation Triangle was formed over three lunar and solar years, using the day-inch unit of time measurement.

Author photographs are from his and his brother's 2010 survey of Le Manio.
See Heath and Heath, "The Origins of Megalithic Astronomy as Found at Le Manio," 2011.

---

*This unit of measure, sometimes called a pygmy foot, was clearly employed in the Parthenon (see figs. 2.6, 4.5, and 5.16).

These could always be reconstituted using the triangular method, providing the English foot, or another known ratio of that foot, was preserved as a singular reference; that is, foot ratios and combinations of them within calculations using ratios to the foot replaced triangles. Just as in writing, there does not need to be speech though there can be reading, geometry could use ratio-based measures in many new ways when comparing and analyzing measurements of length.

This internalization of the triangle was revolutionary in its power to express, within the developing human mind, the angelic science implicit in the design of the geocentric world. In the physical world, there is geometry out there (*geo* means "Earth"), while ratios could become *known to the mind* and applied within a set of transforms available between the different metrological ratios. The right triangle was the stepping-stone to this now unfamiliar form of calculation belonging to the angelic mind, which expresses itself through the recurrences and ratios found within solar systems like ours.

While sacred geometry has since given religious significance to geometrical forms, for example, in Gothic cathedrals, there has been a purist attitude that this should all be achieved through "compass-and-straightedge" geometry, which is proposed to be the case in "the mind of God," that is, for angels. This idea has caused metrology to be suppressed, yet measures are essential to understanding ancient artifacts since they carry the traces of ancient calculation.

The English foot, as the root of metrology, echoed the fact that angelic science is founded on oneness, since there would be no angels unless there was a Universal Will governing creation. Metrology would resemble the cosmic order by being a single, well-ordered whole based on a *single unitary standard,* exactly because this was the form the cosmic order took.

The cosmic design keeps life and consciousness connected with the higher worlds so that the reason for life's existence can be known, which is a common theme in religious thought concerning the power of limits, intelligibility, and harmony between parts. It is within the gift of rational fractions to maintain intelligibility, but numbers on their own are unable to define an ordered universe. Our *lives* are more than just bodies due to their relatedness and movement within a world where ratios relate and transform the physical world.

The angelic world must sit between oneness and physical diversity as the self-consistent power of patterns in eternity (see chapter 9). Religious buildings have come to represent potentials in the eternal world for a spiritual order on Earth, but, to represent those potentials, the world of number ratios was applied through metrology and geometry to make such buildings representative of the eternal pattern. Also, geometrical problems could be calculated and solved using *rational* measures before our historical numeracy was developed starting in Egypt and Mesopotamia. The ordering of geometries on the megalithic landscape and their mastery of time cycles (even using circumpolar stars as a clock),[3] led to new questions such as how large the Earth was.

We don't know whether the angelic minds directly influenced the megalithic astronomers who developed these numerical methods but, as with modern science, progress was rapid.* Ratio-based measures were evolved to design versions of the angelic geometrical forms employed in the forming of the Earth, the Moon, and its planetary environment as seen from the Earth (see "Circles and Squares" in chapter 2).

ANCIENT METROLOGICAL FOOT RATIOS

Since a ratio takes the form of a number (a "multiple"†) divided by another number (a "submultiple"), the *re-measurement* of a measurement by another foot ratio becomes the simultaneous multiplication and division of the measurement. Counterintuitively, the multiple of the foot ratio numerically divides the original measurement, making it smaller, while the submultiple numerically multiplies the original measurement, making it larger.

In the rest of this book, all feet *not attributed* to a named module can be assumed to be English feet. All other foot measurements are *prefixed* to the word *feet,* as in "5000 Roman feet." And the same will be true with aggregate units such as cubits.

So, for example, a measurement of 72 feet can be converted to Roman feet (24/25 feet). Divided by the multiple 24, there are three 24-foot lengths in 72 feet, whereas multiplied by 25, there are three lengths of 25 Roman feet forming 75 Roman feet within 72 feet.

---

*Witness the development of the theories of the atom, quantum mechanics, and relativity in just a few decades around the beginning of the twentieth century.
†This was the terminology of the Pythagoreans, in which an integer number is a "multiple" of 1 and a "submultiple" is its reciprocal, or 1 over the multiple.

When limited to integers, one does not know which foot ratio might divide into a measurement so as to resolve it as an integer. Metrology's method of proportional variation was limited to performing multiplication and division of physical lengths to, if possible, achieve an integer measurement. Without developing the arithmetic of multiplication and division, or the notation of numbers that is necessary for our arithmetic, problems only became soluble using proportionality. Where a measurement is fractional using one measure, another measure can achieve an integer result, and this was especially useful when manipulating the important irrational fractions such as π and the Golden Mean (*phi*).

Therefore, in ancient metrology, measured lengths had *different numerical values* when using different types of foot measure. For example, the English mile is 5280 feet, 8 furlongs of 660 feet, a length widely used for furrowed fields hence the name. But the English furlong is also 600 Saxon feet of 1.1 (11/10) feet, and 5280 feet, the English mile, also 4800 Saxon feet. The standard mile in any foot ratio was 5000 feet, the English mile being 5000 Saxon feet and the Roman mile, of 5000 Roman feet of 24/25 (0.96) feet, was also 4800 English feet. Aggregate units, such as steps (2.5 feet), yards (3 feet), furlongs, and miles were useful as *intermediates* to a large measurement made in any given foot: 100 Roman miles would be 480,000 feet long, yet also 500,000 Roman feet long. In this, one can see aggregate units as similar to the counters of the decimal system where 10 feet plus 100 feet become 110 feet or 100 Saxon feet.

Were one to divide 4800 feet (the Roman mile) by a furlong of 660 feet, the result would not be an integer. To understand why, look at the prime numbers greater than 2 (in bold),* "under the bonnet" of the number 660, which is (4) × **3** × **5** × **11**. This number 660 hides its factors of (4) × **3** × **5** = 60, of which there are 11 in 660. In the number divided, namely 4800, there are no elevens, and so 4800/660 = 7.2<u>7</u>.† The result has a fractional part because 11 is in the denominator of the result 4800/11 = 7.272727. The repeating "27" is caused by the base-10 notation we use, which again has no eleven in its implicit divisions by powers of 10 (2 × 5).

Integer results, which is what rationality originally meant, are only

---

*In each equation, though (4) is 2 × 2, the powers of 2 are not here relevant.
†Repeated fractions such as 0.27272727 . . . are truncated, in this work, by an underlined repeated part, such as 0.<u>27</u>, indicating cyclic repetition.

maintained or achieved when the measures used are commensurate* with a measurement. This draws our attention, from integer ratios per se, to the prime numbers that belong to those ratios.

The prime numbers used within the foot-length ratios of ancient metrology were limited to the set {2 3 5 7 11}.† No ratio containing a number with a higher prime number was allowed into metrology's ratios, just because other prime numbers (such as 19 or 233) would then become visible within a measurement, as significant and, later on, as *sacred* numbers. It was not necessary or desirable for higher primes to clutter up ancient metrology once an accurate ratio of π, which was required for working with circles and squares, had been found to be 22/7 (3.1429), which only needed primes {2 7 11}. In fact, the longest side of each right-angled triangle represents a single point on an invisible circle's radius (see fig. 1.1, p. 13), and many circular structures came to be built using the primes 7 and 11 that are built in to ancient metrology, as with the model of equal perimeter. In chapter 2, "Models of Time and Space," the possible relationships equating the perimeters or areas of squares and circles are identified as being the relationships sought by the angelic mind in defining the Earth's relationship to its Moon and the Moon's relationship to the Earth's orbital year and daily rotation relative to the motion of the Moon's orbital nodes around the Sun's path, the Ecliptic (the model of equal area). This importance of the circle, as typifying the world of eternity, was very influential in the West due to Aristotle. He went too far though in assuming that the planets "in eternity" must have perfectly circular orbits, thinking metaphysics rather than physics would apply directly to the heavenly world.‡

One can therefore state that ancient metrology was all the possible integer ratios developed from the number 1 (the English foot) within the range 9/10 feet and 7/6 feet, a range of 35/27 and only employing the factors

---

*Literally *com* (meaning "inter") and *mensurate* (meaning "measurable"), as whole-number integers.
†The numbers in curly brackets represent a set. This way of enumerating a set of numbers has been found to be better than 3, 4, 5 or "3 4 5" or 3-4-5 and is standard to modern math. In this book, the curly bracket will also be used to designate the set of numbers forming the side lengths of shapes, such as a {3 4 5} triangle.
‡As stated in the article "Classical Astronomy" on the Australia Telescope National Facility website, "According to Aristotelian cosmology it was only within the sub-lunary sphere, that is between the Earth and Moon, that changeable phenomena such as comets could exist."

{2 3 5 7 11} relative to the English foot. Below this range are subunits such as inches and digits, while above are the common aggregates of feet starting with the remen (6/5 = 1.2 feet).

Historic metrology is therefore a backward-looking subject of how different measures were discovered and named during the historical period. Its various units of measure were often given names belonging to the first or most significant location where they were identified, and hence it is a scattered tradition rather than a unified phenomenon. The black foot of the Egyptian Nileometer, for instance, is a kind of Russian foot. John Neal's realization, that each historic measure had a root version rational to the English foot, showed all the measures as having been conceived of *as a whole* and *at a single place and time.*[4] And that metrology was in fact derived from the patterns found within the field of integer numbers, in the order that they arise from the number 1, and then found in the interval ratios containing factors limited to the set {2 3 5 7 11}, used in ancient metrology.

To demonstrate the power of this system, take the Samian foot, named after the Greek island of Samos, birthplace of Pythagoras, the heliocentric astronomer Aristarchus, and ancient historian Herodotus. Herodotus quoted the (southern) 756-foot base length of the Great Pyramid as 800 of his feet, a foot then of 0.945 feet, macrocosmically seen in the 945 whole days found in 32 lunar months (chapter 2). The root Samian foot is 33/35 (0.9429) feet, and it naturally forms a cubit of $33/35 \times 3/2 = 1.4143$, which is very close to the square root of 2 ($\sqrt{2} = 1.414213562$), irrational until such a close integer ratio can accurately approach it. The Samos foot used by Herodotus was 441/440 of the root value, the ratio (a) between the mean Earth and polar radii, and (b) between the Great Pyramid's designed height and its truncated height—as a model of the Earth.

Though against irrationality,* Pythagoreanism could use the geometrical technique called quadrature in monument building by using a Samos cubit so close to $\sqrt{2}$ as to make the diamonds between the squares (doubled or halved in size; see fig. 1.4) integer measurements of this cubit. Intermediate nested diamonds and squares are convenient for doubling or halving perimeters and side

---

*One can see why this is so, given ancient metrology's use of rational fractions to transform irrationals and to perform nonarithmetic multiplications and divisions through remeasuring lengths and building geometries.

lengths: the root Samian cubit and English foot allowed both the squares and diamonds of quadrature* to be integers.

A powerful design tool, quadrature also has a harmonic significance in that the Saturn synod has a square-root relationship to the harmonic model of time (in the octave doubling from 9 to 18 lunar months, see chapter 5), another angelic model that related all the planetary and other time cycles to the lunar year, solar year, and the Earth's rotation—and to make the Earth a suitable geocentric home for life. The diamond is therefore symbolic within quadrature when the square within which it is nested, of side length 18, is halved to 9 lunar months. The lunar year (12) is 4/3 of 9 lunar months while the Saturn synod of 12.8 lunar months is the harmonic tritone ($\sqrt{2}$)[†] of 18, the harmonic ratio 64/45 (1.4<u>2</u>)—as presented in the Canterbury pavement (fig. 7.25, p. 177). To

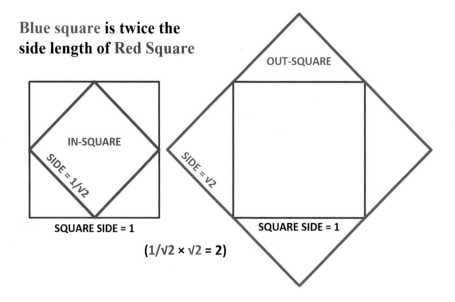

FIGURE 1.4. The technique of ad-quadratum or quadrature

---

*Quadrature exploits the fact that a right triangle with two sides each of length 1 will have a third side the $\sqrt{2}$ (by the law of Pythagoras). If the third side length is used to form two sides of another right triangle, the new third side will be 2 long. The same will happen in getting smaller, and so figure 1.4 shows how the simple act of creating diamonds, as in-squares and out-squares of a given square, naturally forms a unitary pattern of divisions and multiplications by 2 and $\sqrt{2}$.
†$\sqrt{2}$ = 1.4142.

aid in such constructions, the √2 could be accurately modeled using the cubit of the Samian foot (1.4143) or, less accurately, 5/6 of the Royal cubit of 12/7, that is, 10/7 = 1.4286 feet.

Though we can see foot ratios as having a fractional part, these are not irrationals but rather rational fractions: their fractional parts are **finite** or **cyclic** (in base-10) and so can be represented by a rational fraction containing only products of the set {2 3 5 7 11}. More on the development of the metrological system itself can be found in chapter 3, "Measurement of the Earth," further documented in appendix 1.

METROLOGY OF THE FIRST ELEVEN SQUARES

John Michell and John Neal observed an interesting phenomenon in their "W" diagram, in which the many rational modules of metrology seemed to emerge from the squares of the first and smallest numbers.[5] Why this was so emerged from work on the Kaaba (see chapter 8) and the observation that the megalithic astronomers of Carnac used a technique I call "proximation" (pp. 28 and 244), in which two ratios were used to achieve a rational connection between two measurements that would otherwise be irrational. Since *two* small-integer ratios are involved (integers less or equal to 50), the composite ratios can involve larger integers as their multiples and submultiples. Examples from metrology are:

- (22/7 × 8/25) = 1.0057 = **176/175**, but this also equals (24/25 × 22/21). The first pair are approximations to π, whereas the second pair are ancient types of root foot, in feet.
- 63/10 × 7/44 = 1.0023 = **441/440**, but, again, this is also equal to 21/20 × 21/22, and again, the first pair are types of π and the second types of feet.

These show smaller ratios involving three-figure numbers such as 176/175 as emerging from larger ratios like 22/21, examples of the ancient foot ratios and/or approximations to irrationals, such as π as 22/7 or the √2 as 99/70. This can explain the "W" diagram of foot lengths and its relationship to squares that are also seen in the Kaaba, and also how squares form proximate ratio pairs that methodically generate all of the significant ratios found in musical harmony but also the ancient foot ratios and the geometrical and geodetic microvariations applied to foot ratios that constitute ancient metrology.

TABLE 1.2
THE GENERATION OF RATIOS
FROM THE "EARLY" SQUARE NUMBERS

$2^2 =$	4 and	$2/3 \times 2/1 =$	4/3—which is a musical fourth
$3^2 =$	9 and	$3/4 \times 3/2 =$	9/8—which is a musical whole tone and the pygmy foot used in the Parthenon (see chapter 3)
$4^2 =$	16 and	$4/5 \times 4/3 =$	16/15—which is the root reciprocal Samian foot
$5^2 =$	25 and	$5/6 \times 5/4 =$	25/24—which is the Manx foot (inverse of Roman foot)
$6^2 =$	36 and	$6/7 \times 6/5 =$	36/35—which is the root of the common Greek foot
$7^2 =$	49 and	$7/8 \times 7/6 =$	49/48—which is the inverse of the common Egyptian foot
$8^2 =$	64 and	$8/9 \times 8/7 =$	64/63—called here the root Byzantine foot after its canonical variant, found at the Hagia Sophia, of 128/125 feet
$9^2 =$	81 and	$9/10 \times 9/8 =$	81/80—which is a microvariation of the English foot found in the Parthenon (see chapter 3), equal to the synodic comma of musical tuning theory (see chapter 5)
$10^2 =$	100 and	$10/11 \times 10/9 =$	100/99—which is the root reciprocal Byzantine foot
$11^2 =$	121 and	$11/12 \times 11/10 =$	121/120—which could be a rare variation
$12^2 =$	144 and	12/13	—which breaks the limiting set of {2 3 5 7 11}!

The principle behind this list of the first eleven squares is that for each N:

$$N^2 \div (N^2 - 1) = N \div (N + 1) \times N \div (N - 1).$$

For example, where N is 10:

$$10^2 = 100 \text{ and } 10/11 \times 10/9 = 100/99$$

This expresses the fact that a square (N^2) is perfectly symmetrical with itself but that its symmetry can be factored (or broken) by the two superparticular fractions directly adjacent to N in the number field. And this is only one instance of symmetry at work in which, for example, the fractions for an irrational such as π can provide near cancellation (see above examples 176/175 and 441/440), so that employing their sum enables integer results to be maintained between two measurements, where an irrational constant would otherwise demand that one be a non-integer.

This goes to the heart of Neal's *All Done with Mirrors,* in that the mirror for ancient metrology was a *bracketing of the mean.* In simplest terms, the mean measure is 1, from which the foot measures were projected as greater and lesser rational fractions of the English foot. But microvariations such as 176/175 have a mean that is π itself, which can then be seen in its approximation within 176/175, that is, $22/7 \times 8/25$. An integer radius can then generate an integer circumference if that circumference is seen in units that are enlarged by 176/175. The result, accurate but not perfect, does not need to be perfect unless you are flying to the Moon.

MEGALITHIC USE OF ANGELIC RATIOS

Another example, seen at work at Crucuno, near Carnac, was the counting of lunar months using 27 feet, that is, 27 feet equated to 29.53125 days,* which makes each counted day smaller and equal to 32/35 days, the Iberian foot. This allowed days to be counted within the months of 27 feet, that is, 27/29.53125 = 32/35, since $27 \times 35 = 32 \times 29.53125$. (See figure 1.5 on the following page.)

These illustrations only scratch the surface of what appears to be an alternative to arithmetic in that, when you set up the units correctly, one can operate in two worlds, days and months, that normally require fractions of the other unit. Through this, these astronomers also saw celestial time periods, such as the lunar month, had already been made intelligible in this subtle way. Further evidence for this kind of working can be seen at Locmariaquer, east of Carnac and Crucuno, where the end stone of the Table des Marchands dolmen displays a table of numbers as crooked lines facing away from the center, the numbers being {27 32 35} (figs. 1.6, p. 27 and 1.8, p. 29).

In this way, proximate bracketing could exploit numerical differences about a center, using the types of ratio found in ancient metrology. The crooked lines of Table des Marchands are in two columns with their crooks pointing away

---

*This is the value used at Carnac after it was discovered that 32 lunar months were 945 days so that 945/32 = 29.53125 days, which is a *very accurate* approximation.

The Triple Monument of Crucuno
for Counting between Eclipses

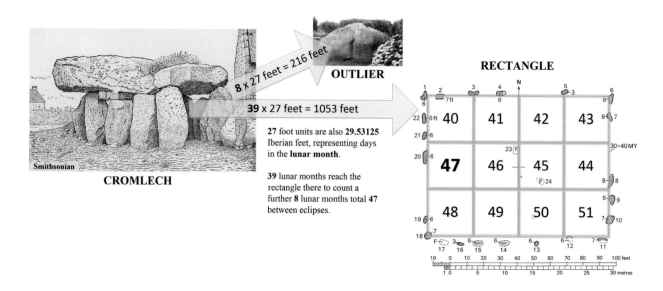

8 x 27 feet = 216 feet

OUTLIER

39 x 27 feet = 1053 feet

27 foot units are also **29.53125** Iberian feet, representing days in the **lunar month**.

39 lunar months reach the rectangle there to count a further 8 lunar months total **47** between eclipses.

CROMLECH

Smithsonian

RECTANGLE

Closeup of the cyan-colored enclosed left-hand side of the lower image showing the distance between the dolmen and outlier Crucuno monuments.

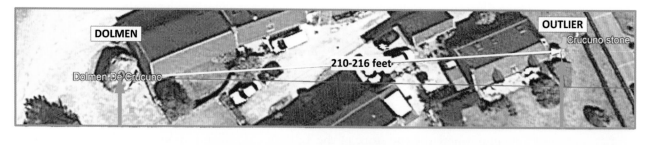

DOLMEN

Dolmen De Crucuno

210-216 feet

OUTLIER

Crucuno stone

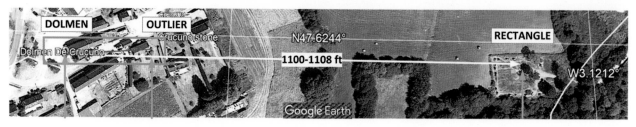

DOLMEN

OUTLIER

Dolmen De Crucuno

Crucuno stone

N47.6244°

1100-1108 ft

RECTANGLE

W3.1212°

Google Earth

FIGURE 1.5. The triple monuments of Crucuno symbolizing the four eclipse periods (*top*) and the monuments and numbers on as seen on Google Earth (*bottom*).

from the center. In each row, the number of lines on the right are one greater than those on the left: implying superparticular ratios. In the case of Crucuno, I suggest the following was true: once the 945 days in 32 lunar months had been identified as an anniversary between a whole number of solar days (945) and a whole number of lunar months (32), 945 days needed to be factorized through discovering which numbers divided into the length, something like figure 1.7.

FIGURE 1.6. End stone from the Table des Marchands dolmen at Locmariaquer (*left*) and a drawing of its weatherworn carved relief (*right*).

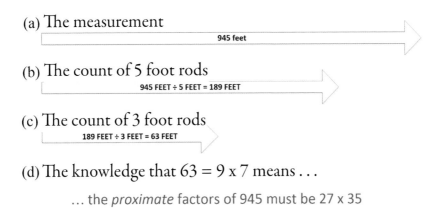

(a) The measurement

945 feet

(b) The count of 5 foot rods

945 FEET ÷ 5 FEET = 189 FEET

(c) The count of 3 foot rods

189 FEET ÷ 3 FEET = 63 FEET

(d) The knowledge that 63 = 9 x 7 means . . .

. . . the *proximate* factors of 945 must be 27 x 35

FIGURE 1.7. Analysis of a 945-foot length for its factors.

Today, we look and see 5 at the end of the number 945 and know 5 must be a factor of it. At Carnac, there was only a day-count *length*—on the ground. How else to store this and manipulate it except by knowing its factors? With a rod 5 feet long, there would be 189 rods in 945 feet with nothing left over. To find a further factor of 189, it is helpful to have made a new length of 189 feet while moving the 5-foot *factorization rod,* allowing further factorization. When 5 feet won't work anymore, a yardstick of 3 feet could be tried, leading to a new count 63 feet long.

Smaller numbers such as 63 would have become familiar to the megalithic specialists, and 63 soon factorized as 9 × 7 and the full factorization of 945 arrived at as 3 × 9 × 5 × 7. We would call its factors $3^3.5.7$ but, the more practical needs of the megalithic culture were to find ratios *proximate* to each other. The cube of 3 is 27 and 5 × 7 is 35, and 945 could then be seen as a rectangle of 27 feet by 35 feet, with an area 945 square feet—since square and rectangular geometries were obviously a relevant way to hold the factors of larger numbers.

Once it is known that 945 days are 32 lunar months, 32 is proximate to 35 and the desired measurement for the lunar month is 945/32 days, which is equal to 27 × 35/32. This leaves 27 feet as a proxy for the lunar month of 29.53125 (945/32) days, while days can be counted as 32/35 feet, the root Iberian foot. In such an early metrology, rational fractions allowed these sorts of transformation of units of measure according to a measurement's factorization and any available proximation pairs, in ratio.

Feet were large enough to support the use of a range of fractions, and a new type of foot was created, in this case to allow the necessary fraction of 35/32. The effect of *dividing* ordinary feet by another rational type, such as 32/25 feet, was to *multiply* by the reciprocal, namely 35/32. This is experimentally obvious: dividing a length by the smaller 32/25 makes the original *measurement,* in feet, a larger *number* of the lesser feet of 32/35. The only other factor is 27, so that when one divides 27 feet by 32/35 feet, the formula of the result is 945/32, the length of the lunar month (29.53125) in the new feet. This is initially hard to understand, but 945 days divided by 32 is 29.53125 days, accurate to 57 seconds in more than every 2.5 million seconds!

There was then no need to actually count the 29.53125 days implicit within the 27-foot month counts, but the resulting two facts sufficed. A day could be measured as a foot of 32/35 feet within each section (or lunar month) when the larger count *used* 27 feet per month. In the early fourth millennium BCE, this discovery would have unified the practices of day and month counting that had formerly been done using either the inch per day or the MY per lunar month.

This is exactly what I found (using Google Earth) within the Crucuno rectangle and then at the Crucuno dolmen, namely that 27-foot units were being used to count, from the dolmen to within the rectangle, as 47 months, a time cycle that functions as the shortest coherent eclipse period, of four eclipse years: the Octon of eight eclipse seasons.

Based on the units obtained at Crucuno, the Table des Marchands appears to show us an interesting table of these factors, as in figure 1.8. And from this one can see how, by the time the Gavrinis chambered cairn was created, 4 kilometers east of this stone, units of length are being demonstrated, on the engraved stones garnered from the many sites to the west, each having weathered

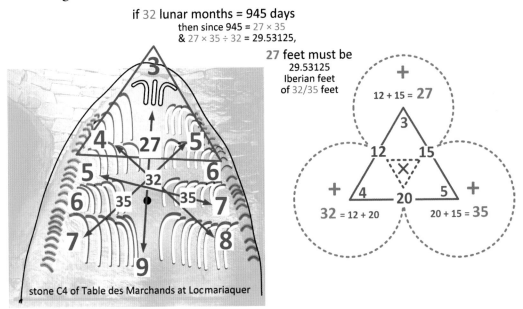

Deriving the lunar month as 945/32 or 29.53125 days

if 32 lunar months = 945 days
then since 945 = 27 × 35
& 27 × 35 ÷ 32 = 29.53125,

27 feet must be
29.53125
Iberian feet
of 32/35 feet

stone C4 of Table des Marchands at Locmariaquer

FIGURE 1.8. The factors 27, 32, and 35 on the end stone of the Table des Marchands.

differently in their original sites (fig. 1.9). These units of measure at Gavrinis and Crucuno firmly belong to the system we now call ancient metrology: the foot and Royal cubit were being shown before 3200 BCE, that is, before Egypt would use that measure and, from then on, all monuments worldwide employed *the same system* we aim to understand here. We have seen how the fundamental unit, the English foot, emerged *automatically* when the Lunation Triangle was reused to count months instead of days, using MYs to represent the lunar month. And I have found proximate measures were developing and being used within the monuments of Carnac in order to resolve astronomical time cycles and ratios. It seems number ratios, alongside geometrical methods, were the natural teachers of these astronomers and that both were necessary because

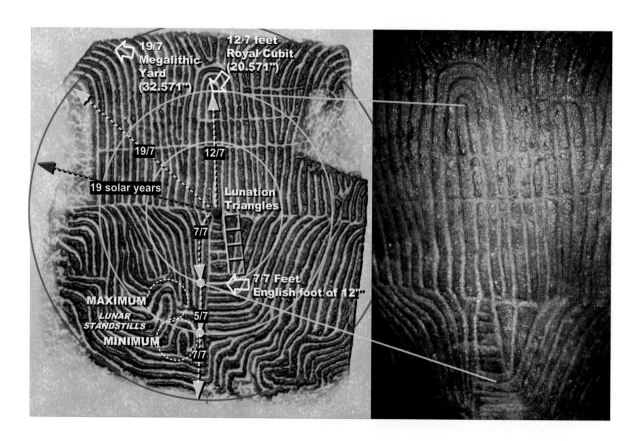

FIGURE 1.9. Stone C3 at Gavrinis containing the English foot, Royal cubit, and AMY.

Image on the left courtesy of Cassen, Lescop, and Grimaud with 3-D modification by David Blake; image on the right DuVersity.org, taken by the author.

they undertook their task before the advent of arithmetic and were fortunate in being able to follow the methods of the angelic mind in this way.

Stone C3 (fig. 1.9) shows the three measures (foot, Royal cubit, and megalithic yard) evolved through the successive re-use of the Lunation Triangle geometry: starting with a day-inch count giving the megalithic yard as the excess over the lunar year; then again, counting in lunar months in a single solar year giving the English foot of twelve inches as excess.

1. Stone C3 is approximately circular; it is 19/7 feet in radius from its center.
2. Below the center point, what looks like a menhir is divided into notionally equal portions.
3. Seven of those portions equal the English foot so that each portion is then 12 inches divided by 7, or 12/7 inches. That is, the lines on the engraved stones of Gavrinis show divisions that range between one inch and 12/7 inches, and in this case the latter.

In feet, the ratio 12/7 is the Royal cubit—here displayed long before the Egyptian use of that measure (yellow circle). Adding the portion below this "menhir," to the foot, gives 8/7 feet—the Royal foot—and the portions below end after twelve so that the Royal cubit is shown as a radius $12 \times 12/7$ inches $= 144/7 = 20.\underline{571428}$ inches (blue circle). The extent of the stone being 19/7 feet (purple circle), the whole tableau evidently dealt with these three septenary* units, most famous in historical metrology, yet here before 3000 BCE and hence prehistorical. The line above the center of C3 is also 12/7 feet and, adding 7/7, the English foot, gives 19/7 as the central meaning of stone C3. Moonrise at maximum standstill in the south illuminates the whole chamber, but the Sun at the winter solstice is shuttered so as to only shine on the left side of stone C3.

Dividing the English foot into seven parts is difficult without using the right triangle: seven parallel threads 1 inch apart could form a base above which a foot of 12 inches would have the required unit, used between the engraved lines, between those threads—in an extension of the Neolithic weaving

_____
*Septenary means "relating to or dividing by seven."

technology. This was a technical language explaining astronomy using geometry and metrology, and the collected stones of Carnac appear part of the lost intellectual content of the megalithic. In Britain they became badly weathered outdoors and so became unintelligible, while in the Irish chambered tombs engravings remarkably like those of Gavrinis still also survive. Engraved pictures would have to be well planned, with their key measures established between points and these then joined with the distinctive, parallel yet curvilinear lines.

There are two "carrots" beside the "menhir" in figure 2.8. These are congruent with the Lunation Triangle and hence, of half a four-square rectangle, which is the incredibly simple method by which that triangle could be reproduced to high accuracy without doing any counting, using a base length of 12 megalithic yards to generate the English foot. Stone C3 therefore expresses the key discovery process that gave the megalithic access to the world of astronomical time so as to communicate it to others given the triad of monument, stone art, and spoken word. Without the spoken word, access to modern astronomical data has allowed the monuments, and in some cases stone art, to tell the story of why and how they were made.

2

Models of Time and Space

ANCIENT SITES, AND THE ICONOGRAPHIC and textual works of the ancient world, can now be reinterpreted by recovering the ancient system of measures used, deducing why geometrical forms were employed, and understanding their well-developed musical tuning theory. These together have reconstituted what can only have been models and techniques naturally present in the angelic mind, themselves reconstituted by megalithic astronomers in their building of monuments to re-express the cosmic work of the angels. Perhaps this is why these buildings often have a religious connotation or a ritual purpose, and we must assume that the work of building them was done to reveal how the Earth and its cosmic environs had been finely tuned. We now know that life and eventually intelligent life is quite special among planets, but, in the megalithic, the details of how special the Earth was continued to appear. The resulting humans discovered the astronomical models of cosmic architects, and left these models upon the Earth, models describing the relative size of the Earth and the Moon, the Earth's shape or geoid, and the harmonic relationship of that Earth to the geocentric time-world of all the other planets.

As already stated, this sharing of "mind-stuff" with angels developed the human mind's capacity for rational thinking and numeracy, which fed into the civilizations that are largely responsible for monuments that expressed the angelic models and the techniques for arriving at them. The models are geometrical, and the numbers are subliminal within geometry, as their form often leaves their dimensions unquantified or, indeed, without any scale or frame of application. The interactions of measures within space—that is, within geometry—are best expressed in the right-angled triangle, more simply called a right triangle: a powerful geometrical model.

RIGHT TRIANGLES

The right triangle is remarkable, as the Pythagorean triangles* show. Right triangles are an interface between pure number and the world of space and, more generally, they incarnate the trigonometrical functions we know as sine, cosine, and tangent since they represent a point on a circle (fig. 2.1) of a radius equal to their longest side.

The First Triangle

The First Triangle of Pythagoras involves sides of lengths 3, 4, and 5, and it is more than remarkable that these very small numbers *can* come together within the strict discipline that the sides must be integers and one of the angles a right angle. As a consequence, it is possible for any person interested in playing with twelve identical objects, such as cubes or dice, to see that a 4-by-3 rectangle made of these objects has a diagonal that is 5 long (fig. 2.1).

The relationship of numbers to the spatial dimensions is revealed to the mind by the {3 4 5} triangle. The harmonic relationships of the *senarius* {1 2 3 4 5 6} (see the discussion of harmonic models in chapter 5) are found in its lengths as ratios of 4/3, 5/4, and 6/5, implying that the domain of harmony

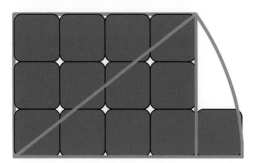

FIGURE 2.1. Twelve dice as "3 and 4 mated with 5," the First Triangle of Pythagoras, Plato, and dynastic Egypt. The diagonal can be arced down, as shown, showing its length to be 5 units.

---

*Pythagoras (c. 600 BCE) founded a long-lasting tradition of geometrical number science based upon the Mystery teachings. This tradition gave special attention to triangles with one right angle and its three side lengths simple integers. These "Pythagorean" triangles and his famous law, about the squares of the sides of a right triangle, were known to "the Mysteries" since the fourth millennium BCE, in Egypt and Babylon.

was built into the world of space as a framework condition. Also interesting is that 3 × 4 equals 12, introducing what would become known as an area, itself contained within the idea of volume. Volume impressed the ancient world with numerically designed vessels that could contain so many grains of a crop as a "grain standard," which then became a way of visualizing large numbers. A three-dimensional vessel with sides of {3 4 5} would hold a volume of 60 units, the base chosen by the Sumerians circa 3000 BCE. There is a subtle difference between this numeracy, which seeks to understand the world of space and number, and our own, which seeks to solve problems in the physical world. This difference has made modern numeracy blind to the significance of why numbers are related to the properties of space.

The astronomers of Carnac, in northwestern France's Brittany region, have left testament to their use of that location's latitude (42.5 degrees north, rather than any other) where alignment to the extreme solstice Sun, in winter and summer, is aligned to the 5-side of a {3 4 5} triangle. Instead of waiting for the

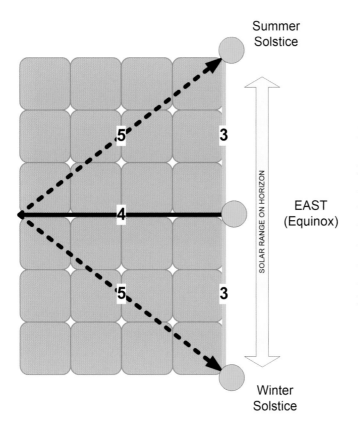

FIGURE 2.2. Megalithic Carnac's choice of that latitude showed the solstice Sun aligned to the {3 4 5} triangle. The five sides to the solar extremes would be observed from any observing site around Carnac, at the common vertex, here on the left (the west).

extreme Sun to occur, the longest side of a {3 4 5} triangle aligned to the cardinal directions (fig. 2.2) could prefigure it. Counting between these extremes gave a count of half the solar year, and a whole cycle to the first solstice, counted in days, gave the solar year (365 days).

The Lunation Triangle

In terms of our focus here on {3 4 5} triangles, three rectangles (of twelve dice each and composed of 3 vertical dice by 4 horizontal dice, as shown in fig. 2.1), can be joined together to make a 12-by-3 rectangle. This {3 12} rectangle is of great consequence for the world of time on Earth: its diagonal is in proportion to its base as the solar year is to the lunar year of 12 lunar months. The implied triangle of base {12}, diagonal ($\sqrt{153}$ = 12.368) and third side {3} is the Lunation Triangle* because it harmonizes the otherwise irreconcilable lunar and solar calendars. Also remarkable is that three {3 4} rectangles, each containing the first {3 4 5} Pythagorean triangle, should relate this most crucial pair of time periods between the Sun, which creates almost all the light on the Earth,

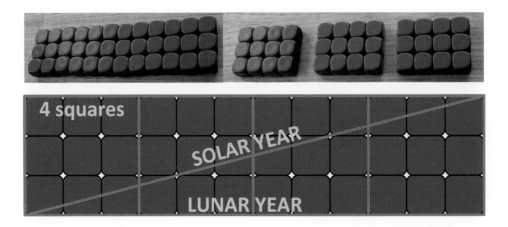

FIGURE 2.3. Twin triangles within the four-square rectangle. These triangles very accurately define the ratio between the solar year and the lunar year of twelve lunar months. It provided early astronomers with an easily constructed framework for reproducing the Lunation Triangle.

---

*Called the Lunation Triangle by Robin Heath (1993 & 1998), since it naturally models the lunar year as its base length and gives the solar year in lunar months as its longest side ($\sqrt{153}$ = 12.369). During a survey of Le Manio in 2010, it became obvious that the third side-length {3} meant the Lunation Triangle was half of a four-square rectangle, revealing an easy way to construct one.

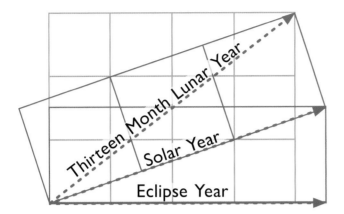

FIGURE 2.4. A three-square rectangle (the Third Triangle) can relate the solar year to the eclipse year. When two are stacked, they span to equal the 3, 4, 5 triangle that, at Carnac, was used to locate the solar extremes relative to the east-west orientation.

and the Moon, which reflects some of the light it has received from the Sun back to the Earth. Early humans were therefore offered a lesson in the geometrical ideality of the Sun and Moon using "sacred" numbers.

The geometry of the four-square rectangle (fig. 2.3) contains two of the Lunation Triangles,[1] a triangle that Carnac's astronomers had realized at the Le Manio Quadrilateral[2] using day-inch counting (see box, "Counting Time as Length" and the section after next, "The Second Triangle"). The maximum and minimum extremes of the Moon at Carnac, over 18.618 years, were very clearly aligned as the diagonals of a single square and a double square, just as the 3-by-4 rectangle gave the Sun's extremes in the solar year at Carnac on the horizon. They would therefore have noticed how the 12-by-3 rectangle (four squares) contains this triangle. The relationship between the solar year and eclipses anywhere on Earth are well modeled by the three-square triangle, giving both the relation of the eclipse year to the solar year as being the same as the solar year to the 13-lunar-month year (384 days), which would otherwise have been much more complicated to realize.[3]

The Second Triangle (and Rectangle)

The number 12 is also associated with the Second Triangle* (also Pythagorean), whose sides are {5 12 13}. This triangle is often found at ancient sites. For

---

*The Second Triangle is the second-most simple Pythagorean triangle. Two such triangles, one opposed the other, with shared hypotenuse for the Second Rectangle, are found in rectangular areas of a city (chapter 5) and notably, the Station Stone rectangle of Stonehenge.

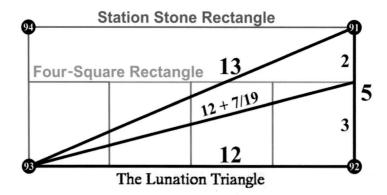

FIGURE 2.5. Stonehenge's Station Stones describe a rectangle {5 12}
whose diagonal {13} forms a Second Triangle {5 12 13}
with a side length {12} which can make a Lunation
Triangle {3 12} if an intermediate hypotenuse is
formed to the 3:2 point of its {5} side.*

example, Jerusalem's old city[4] (fig. 5.15, p. 125) and Teotihuacan (fig. 5.17, p. 130) appear contained by it, and the {5 12} Station Stone rectangle of Stonehenge (fig. 2.5) contained two nested {5 12 13} triangles.[†] In addition, the Second Triangle's rectangle contains the Lunation Triangle and its rectangle of four squares. The Parthenon's cella (or "room") seems to describe it as well (fig. 2.6).

Unfortunately, Carnac in 5000 BCE and Stonehenge in 3000 BCE *are not properly related* by an archaeology that does not recognize the role of numbers within megalithic locations and their constructions because the logical conclusions flowing from that would be unacceptable: that the megaliths were highly technical achievements rather than the creations of a primitive and superstitious race.

---

*As will all triangles, including the {3 4 5} triangle, when joined at their hypotenuse to form the diagonal of a rectangle.

†Robin Heath originally thought, during 1991–96, that the Second Triangle was the geometrical way megalithic astronomers first found the Lunation Triangle, but our joint work at Le Manio shows that the {3 12 √153} Lunation Triangle emerged through day-inch counting yet would have been known to lie within the Second Triangle and hence Station Stone rectangle.

Time Counting, a.k.a. Day-Inch Counting

Time counting using equal units of length, such as inches, was used to represent equal units of time, such as days. This enabled reliable measurements of time between the sky events that delineate time cycles like the solar year (between equinoxes) and lunar years (the twelfth full moon after a full moon). The idea of time counting has been largely overlooked* though many monuments have been found having lengths corresponding to time periods but these are treated as symbolic of the time period, without then questioning where the original data came from and how was it stored by a pre-numerate culture. But if lengths of time were counted directly from day-to-day, then lengths corresponding to time were being directly generated without any intermediaries such as data (a count) written down as a number rather than a length.

*An exception being Alexander Marshack whose *Roots of Civilisation* (1972) found time-factoring within Stone Age art, on bones in particular.

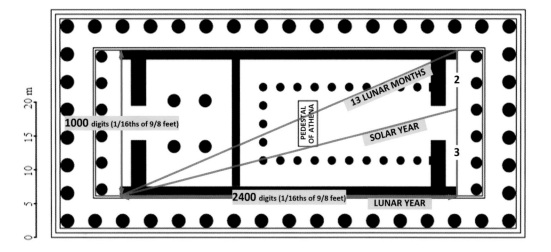

FIGURE 2.6. The Second Triangle in the Parthenon (*plan view*), giving its cella an astronomical context.

Apart from the spatial operations within monuments, the use of triangles in the megalithic according to time counting (see box on p. 39) also identified the harmonic model of the outer and inner planets relative to the Moon (see chapter 5), and this set in motion the formation of a metrology based on an angelic science of ratios (see chapter 1), since astronomy pointed toward both ratios in time and ratios in the geometry of space. One sees an irony in latter-day sacred geometry being based on only "the compass and straightedge", making it blind to the significance of both time and space as being measurable lengths in the megalithic period.

A foretaste of what was then possible is found as the Second Triangle within the Parthenon, where it was found that a foot length of 9/8 feet was used to build its cella as digits, 16 per foot, then 150 such feet (168.75 feet) equals 2400 digits, while the width is 1000 digits. Chapter 3 will explore the geodetic meaning of the Parthenon (see figs. 4.5, p. 81) and chapter 5 its harmonic meaning (fig. 5.16, p. 127), while figure 2.6 shows that the Second Triangle was used to define the cella. (See previous page.)

CIRCLES AND SQUARES

Doubling Using Squares and Circles

The subject we now call sacred geometry had practical roots during its early developmental history. It would have been a way of thinking and a type of language. Experiment and learning were happening in a world without preexisting rules or the preconceptions of later epochs. Sacred geometry is not functionally useful unless you want to build a sacred monument, whereupon it is placed in the category of Being rather than of Will: sacred geometry just is and "does" what it is. In clear contrast, applied mathematics involving geometry, such as of vectors, can model the effects of gravity and say where a ball once thrown will land—meaning it expresses Function. Angelic geometry combined both these aspects in an act of Will (defined on p. 212). The acts of geometry considered here were enduring acts of Will that could explain the world, without having or using our Function-based physics. The only human capacity available for such a thing was the visual imagination, an imagination shared by the angelic world, which can make geometrical and numerical representations out of phenomena.

Intelligence needs a mind in order to act, and planetary intelligence must create minds to realize itself since intelligence is unlikely to be able to construct things without a mind that knows how. Where and how intelligence and mind arose is mysterious because they are of Being and Will, domains only implicit to humans, who are unable to see them directly and can only infer them from phenomena (see chapter 9). This is my way of saying that the geometrical models of this chapter, while part of sacred geometry, were more importantly involved in the Sun-Moon-Earth's development as a system. They are relics from before the invention of sacredness as a term, but they are often but unaccountably present throughout the built heritage of sacred buildings. And we are the minds the Earth has developed, but recently our culture decided to move away from admitting that higher minds than our own first conceived the geometrical models built into sacred buildings.

The Squaring of the Circle by Perimeter Length

Many circular monuments of the last 2400 years, such as domes, appear to exploit a very simple geometrical property, visible as two concentric circles whose diameters, in the same units, are 11 and 14 units.* The reason for this is that the out-square for the 11-long diameter is then $4 \times 11 = 44$ units in perimeter,

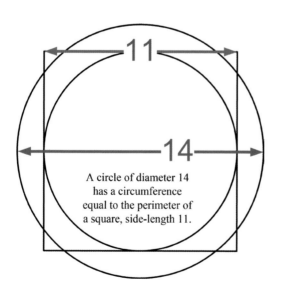

A circle of diameter 14
has a circumference
equal to the perimeter of
a square, side-length 11.

FIGURE 2.7. Visual appearance of the Equal Perimeter Model.

---

*This geometry was crucially rediscovered by John Michell in the early 1970s, through his work on Stonehenge and the Great Pyramid of Giza, monuments that both express it. His influences in its discovery are listed in chapter seven.

while the circumference of the circle 14 in diameter is $14 \times 22/7 = 44$ units. The out-square and the outer circle are therefore of the same perimeter length as the square when π is taken to be 22/7, as in figure 2.7. This arrangement is one type of squaring of the (outer) circle by the out-square of the inner circle—our Equal Perimeter Model (fig. 2.7).

That a circle of diameter 14 has a circumference equal to the perimeter of a square of side length 11 is the key point (connected to π as 22/7), and the initial circle of diameter 11 might be considered irrelevant, but for the fact that the relative sizes of the Earth and the Moon appear to be accurately shown by this geometrical model (fig. 2.8).

The difference between 14 and 11 is 3 units, which can be seen to be a circle of that diameter whose center may be placed anywhere on the outer 14-diameter circle and touch the inner 11-diameter circle. The relative sizes of the Earth and Moon are accurately 11 to 3, and so the inner circle of 11 is the mean size of the Earth and the small 3-diameter circle the mean size of the Moon. The Earth has an equatorial diameter larger than its polar diameter, so that the mean Earth diameter is in between, being the diameter of a spherical planet that does not rotate. We shall see that the mean Earth was considered its spiritual nature and reflected in the perimeters of sacred spaces, as scale models of the mean Earth.

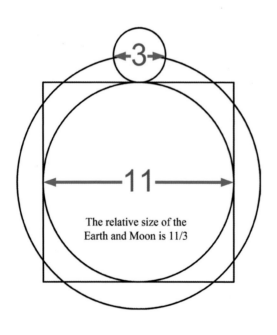

The relative size of the
Earth and Moon is 11/3

FIGURE 2.8. The relative sizes of the Earth and the Moon accurately follow the dimensionality of the Equal Perimeter Model.

The Moon is only seen to touch the Earth at moonrise and moonset, when the Earth is forming the horizon to east or west. Notionally then, the 14-diameter circle could have been seen as a link between the horizontal perimeter of the Earth (11) and Moon (3), where the Moon was seen in the 3 to 11 proportion to the Earth. But this is to think of human observation, whereas the Equal Perimeter Model appears to have been an angelic precognition of how the Earth and the Moon were to be related.

We know that the traditional symbol of the Earth is a square and the traditional symbol of the celestial world a circle. The square is a symbol of the solidity of objects on the Earth, and the apparent flatness of its sides resembles the generally flat surface of the Earth underfoot, especially when extended to a flat horizon of zero altitude. The stars, planets, and Sun travel around the Earth at an apparently equal radius, and the numbers of this model arise from the cosmos, from the outside in. The outer radius is 14, with the natural diameter regulated by the ratio π of 22/7. A diameter 14 has a radius 7, and the outer circumference is 44. Each quadrant is therefore 11 because a square has four sides, so the square of equal perimeter needs a side length of 11, making its perimeter 44. This model of a square of equal perimeter to the outer circle is primary, and only then can the circle of diameter 11, the square's in-circle, be added. But then, what is its circumference but $11 \times 22/7$, or $2 \times 11^2/7$.

The square's perimeter of $44 - 11 \times 22/7 = 242/7$, which leaves 66/7, which, divided by 22/7, arrives at 3 units, the diameter of the Moon. It appears therefore that:

> The circumference of the Moon is the difference between the perimeter of
> the out-square of the mean earth, and the circumference of the mean earth.[5]

It is extraordinary that the Earth-Moon system arose through an early event in which the proto-Earth collided with a Mars-sized protoplanet (now called Thera) and that, in an uncanny act of cosmic accounting, the resulting system had the combined circumferences equal to the out-square of the mean Earth and the relative diameters of 11 to 3. The ratio 11 to 14 of square and circle of equal perimeters and the equal difference between (a) the circumference of both circles and (b) the difference between the perimeter of an 11-diameter circumference and its out-square are simultaneous properties achieved when π is 22/7.

The low numbers involved make the geometry simple, while the Earth and the Moon, having not changed significantly in volume since the collision that caused them, achieved this 11 to 3 configuration billions of years ago. And megalithic astronomy, the earliest culture with the known level of numeracy to infer it, led to the symbolic assumption that the outer 3-diameter circle was the Moon and the 11-diameter circle was the Earth's mean size.

This geometry, recovered by John Michell in the 1970s, is to be found in monuments over a wide geographical range and of different epochs (fig. 2.9). Many pyramids, domes (chapter 4), artworks, church pavements, the Kaaba (chapter 8) and Buddhist mandalas and stupas seem to follow it as a canon. It has a fabulous internal structure, in the proportions between its lengths and in its profound expression of factorial numbers (see chapters 3 and 5).

That this model provides the relative proportion between the Earth and the Moon and their diameters in miles, when multiplied by $6! = 720$, is probably the reason why it became a primary geometrical model for monumental works.* By adopting it, a monument became a scale model of the Earth-Moon system. The ancient metrological system became extended within it, from a tool for counting astronomical time into a geodetic system suited to measuring the size of the Earth more accurately. The Equal Perimeter Model was only an approximation that was supplemented by an improved model of the size of the Earth, which is implicit in the ratios of metrology such as 176/175, 441/440, 126/125 and 3168/3125. This *metrological model* of the size of the Earth will be compared with the 11/3 model in chapter 5, since the harmonic model connects the two through the senarius of {1 2 3 4 5 6}. That model is explored in chapter 3 and is demonstrated by the Pantheon of Rome (fig. 3.5).

The Squaring of the Circle in Area

A well-known challenge to the geometers of the ancient world was to square the circle in terms of its area, so that the square formed should have the same area as the circle. In the early 1800s, this problem was proven impossible to do through geometrical construction, yet the 18.618-year rotation of the lunar nodes around the Ecliptic appears to have achieved this relative to the orbital

---

*The exclamation point following a number indicates the factorial of that number, that is, the number multiplied by every number below it. So $6! = 1 \times 2 \times 3 \times 4 \times 5 \times 6 = 720$.

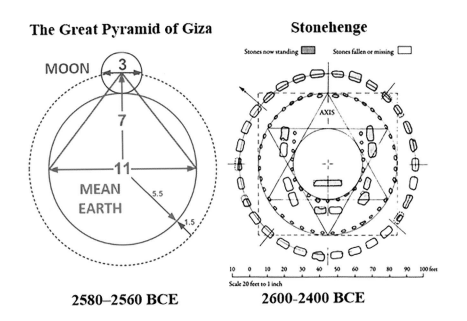

The Great Pyramid of Giza

MOON

3

7

11

MEAN
EARTH

5.5

1.5

2580–2560 BCE

Stonehenge

Stones now standing ▨   Stones fallen or missing ▢

AXIS

10   0   10   20   30   40   50   60   70   80   90   100 feet
Scale 20 feet to 1 inch

2600–2400 BCE

Dendera Zodiac

c. 50 BCE

Aztec Stone of the Five Ages

c. 1500 CE

FIGURE 2.9. A collection of Equal Perimeter Models:
the model itself and the Great Pyramid of Giza, Stonehenge,
the Dendera Zodiac, and the Aztec Stone of the Five Ages.

The image of Stonehenge is from John Michell, The Dimensions of Paradise, *2008, fig. 6.*

period of the Earth (the solar year) and its own rotation every sidereal day. The solar year of 365.2422 solar days is due to the rotations of the Earth in a single year, and the little bit extra of 0.2422 days, close to a quarter, leads to an extra day every four years of 365 whole days in our calendar: every fourth year, our leap year adds a day since 0.2422 × 4 equals 0.9688 days. But this is not exact, and it is only after 33 years that a full 8 days leaves almost no fractional part, 33 years being 12,052.9926 days (that is 12,053 days).

The ratio between 18.618 and 33 as numbers in years is not as relevant as the day count of these two periods. Practically speaking, these periodicities in years are accurately 18.618 years equals 6800 days and, 33 years equals 12053 days. The fraction 12053/6800 is 709/400 (if the *common factor* of 17 is removed from top and bottom of the fraction) and 709/ 400, in our notation, is 1.7725. This squared is 3.14175 . . . , which is π (3.14159) to very high accuracy, nearly one part in 20,000!

This is tantamount to the fact that a circle of radius 18.618 would have an area (= $\pi \times 18.618^2$) equal to the area of a square of side 33 (33^2). Figure 2.10 shows this as the Equal Area Model in years, and this is the only low-number pair able to achieve this, if the square side is to be an integer. The solar year and nodal period are therefore a celestial solution to the "squaring of the circle" in area, normalizing the Earth day with the nodal motion that moves 1/18.618 DAYs in angle* in each terrestrial day. Because of this, after 18.618 years the lunar nodes return to the same place on the Ecliptic, though they remain two invisible points opposite each other in the Zodiac. The nodes can only be inferred through observing their motion between eclipses or as the 18.618-year cycle between maximum or minimum standstill, to north or south on the horizon.†

What happens in *years* of nodal action is due to what is then happening in *days*: if 18.618 days (the node day) are the radius in fig. 2.10, then the area is π times the eclipse year (= $18.618)^2$ days, or π times 346.62 days which equals 1089 days; and this is also the area of 33^2 days as a square of side length 33 days. That is, what is true outside the year (18.618 and 33 years) is

---

*A DAY in angle (shown in capital letters) is the average distance the Sun moves in the starry sky by 1 DAY in 1 terrestrial day. This unit is useful in preserving numeracy between cycles that take years and the motions within the year that have the same numerical value in days. There are 364.2422 DAYS in the Ecliptic (or Sun's path), though it is normally calibrated as 360 degrees in the Ecliptic.
†Or the 9.309 years between maximum and minimum standstill or vice versa.

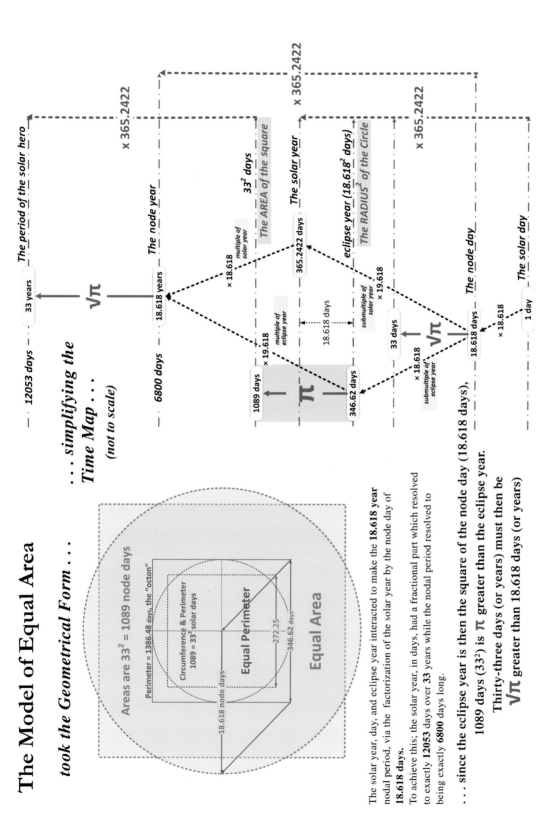

FIGURE 2.10. A circle with radius 18.618 years will be equal in area to a square of side 33 years. This expresses the unique ability of the numbers 18.618 and 33 to equal 1.7725, which is $\sqrt{\pi}$, so that their squares equal π.

also true in days because, the number of days in a node day (18.618 days) is numerically identical to the number of years in a nodal year (18.618 years). These proportions to 33 days and 33 years will then *automatically* stay the same, there being 365.2422 periods of 33 days in 33 years and 365,2422 node days in the nodal period. This parallelism between numbers is shown on the right of fig. 2.10 where the time map shows the eclipse year and solar year are symmetrically held between the 18.618 days of the node day and the 18.618 years of the nodal period. The yellow area shows the ratio of π^* (= 3.14177) between 1089 days (the square of 33 days) and the eclipse year of 18.618^2 (= 364.62 days), *because* 33/18.618 is $\sqrt{\pi}$.

The maximum moonrise in the east (or moonset to the west) is a rare but completely reliable occurrence every 18.618 years—a period appearing in the angelic mind because this *squared,* and multiplied by π, arrives at the perfect recurrence of the Sun every 33 years (a time period associated with solar heroes and especially in the age of Jesus at his crucifixion and rebirth.)

THE THIRD TRIANGLE AND THE GOLDEN MEAN

The Equal Area Model is evidently facilitated in practice by the application of π (of the circle) and of the Golden Mean, also called *phi* (1.618034). This is neither an integer nor geometrical solution, but rather, it is an accurate geometrical approximation involving these two irrational fractions. *Phi* (1.61803399 . . .) can be approximated by the numbers of the Fibonacci series[†], but its decimal fractional part can be truncated quite accurately to being 618 thousandths which we see here in 18.618 days and years; π is delivered with the model by the circle's radius relative to its perimeter since the geometry of a circle uses π's irrational value, though approximations can also be assumed as with the 22/7 of the Equal Perimeter model.

As stated above, the number of years taken by each of the two (diametrically opposite) lunar nodes to traverse the Sun's Ecliptic path is 18.618 years. The number 18.618 is defined by the Moon's nodal period but then per-

---

[*]π is actually 3.14159, to five significant places.
[†]This series grows by adding two adjacent integers to form the next member of the series. Fibbonacci started his series with {1 1} and produced {1 1 2 3 5 8 13 21 34 55 89 144 233 . . .}. These adjacent integers converge to the Golden Mean but can never reach it since it is irrational.

colates into the sublunary world of time, penetrating the solar year purely because there are 365.2422 days in that year, so that the lunar nodes must move 365.2422 days/18.618 days on the Ecliptic in a single year*, which is 19.618 DAYS (in angle) per year. The solar year is therefore 19.618 node days† long and also 19.618 × 18.618 solar days long. That is, *the solar year is factored by the node day.* The eclipse year is short of the solar year by 1 node day so that it is 18.618 node days long and 18.618 × 18.618 solar days long (equal to 18.618^2 or 346.62 days).

Through this cosmic arrangement, the solar day, node day, eclipse year, and solar year are all synchronized to the nodal cycle of 18.618 years. The node day is 1/**19.618** of the solar year and 1/**18.618** of the eclipse year, while the nodal period is both **18.618** solar years and **19.618** eclipse years. We will see a practical example later in this chapter when fig. 2.19 shows the triangular harmonization of the 19-year Saros and Metonic periods as a Golden Mean solution between 19.618 eclipse years and 18.618 solar years. This ratio of 18.618:19.618 is the normalized form of what we will call the Third Triangle (fig. 2.11, *right*) in which the two longer sides of the triangle are 18.618 and 19.618, different by just one unit.

Equivalent Forms of the Third Triangle

The fractional numbers 18.618 and 19.618 needed to be transformed into integers, by a culture employing metrological lengths to hold such number values as integers. A near-Pythagorean triangle can provide just this, by using an appropriately simple number for the hypotenuse, of 19 years—the Metonic period—and a shortest side of 6 years so as to force the base to be only-slightly larger than 18 solar years, in fact, 18.0278 years, while the Saros eclipse period of 18.031 solar years (19 eclipse years) is then only 1 day greater in length

---

*The DAY (in capitals) is the measure of angle reflecting the fact that days were counted by the megalithic culture, and, in this case, these are days of solar movement on the ecliptic, not 360 degrees we use but 365.2422 DAYS in angle. The *number* of solar days taken for a celestial body to move a single DAY on the ecliptic, is reflected as the *number* of solar years the planet takes, to return to the same part of the Zodiac, as seen from the Earth. For example, the Sun moves one DAY in a day and so takes a solar year to travel a YEAR in angle. If the nodes take 18.618 days to move a DAY then the nodal period is 18.618 years long to move a YEAR in angle.
†The node day is the time taken (18.618 days) for the (diametrically opposed) crossing points of the Moon's orbit and the Sun's Ecliptic path to move by a single DAY in angle on the Ecliptic.

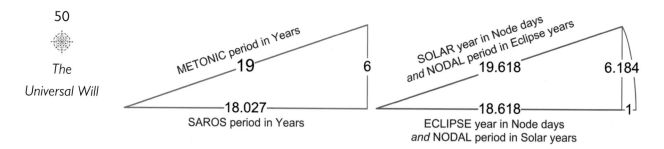

FIGURE 2.11. Near-Pythagorean (*left*) and normalized (*right*) triangles relating solar and eclipse years.

(fig. 2.11). While extraordinary, this is still not a convenient construction, being a relationship in years and involving one fractional length.

However, a pragmatic approach appears to have been used in the megalithic period. Previously, counting periods in terms of the lunar month offered a simplicity we have associated with the angelic mind, and the Saros period is exactly 223 lunar months long while the Metonic period is 235 lunar months long. The solution was to build a right triangle with a 19 long side and a 6 short side on the ground and then count 223 months along the bearing of the base and 235 months along the bearing of the hypotenuse, a bearing by the Third Triangle (see fig. 2.14, p. 53). The Saros period of 223 months is 19 eclipse years long, while the Metonic period of 235 months is 19 solar years long. As the megalithic astronomers of Carnac had found out:

A: multiple-squares geometrically defining significant alignments of the Sun and the Moon at Carnac, and

B: the four-square rectangle defined the relationship of solar to lunar years.

This made the three-square very useful for modeling, visualizing, and symbolizing the eclipse year to the solar year relationship (fig. 2.12).

All this can be seen in the famous alignments at Le Ménec and the less-famous alignment at Erdevan, both near Carnac. The Ménec alignments were set to follow the 18.44-degree angle (of the three-square triangle's diagonal), north of east, because the location's stone rows recorded actual measurements of the Moon's *deviation* from the Ecliptic, on the horizon, sampled every

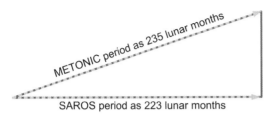

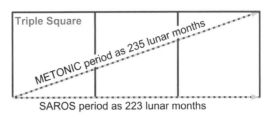

FIGURE 2.12. The simplifying strategies. *Left:* counting the lunar months in the Saros and Metonic periods to form a third similar triangle. *Right:* the congruent geometry of the triple square can represent and access the same ratio, capturing the relative time lengths in a fourth form of the Third Triangle.

82 day-inches* (3 lunar orbits are 27.32166 × 3 = 81.965 days, that is, 82 day-inches) over the nodal period (fig. 2.13, p. 52). Therefore, the monument incorporated the three-square triangle within these alignments, and other stones at appear to show the three-square geometry at the cromlech on their start to the west.[6]

In contrast, the bearing of the alignments near Erdevan (north of Carnac) was that of the imperfect Pythagorean {6 18 19} triangle but, to the southeast rather than the northeast. Both recorded the *relationship* between the solar and eclipse years, but each had a different focus: at Erdevan, the focus was on the 19-year periodicities, rather than on moonrises on the horizon during the nodal cycle, as at Carnac.

This Erdevan project (fig. 2.14, p. 53) probably used the Metonic period as consisting of five of the 47-month-long Octon eclipse periods of 4 eclipse years (as at Crucuno, see chapter 1). Five periods of 47 lunar months (235 lunar months) equal the 19 solar years of the Metonic period,[†] which were then partnered with the 19 eclipse years of the Saros period (223 lunar months).

At Erdevan, the angle to the tumulus of Mané Groh (at 18.4 degrees south of east; fig. 2.15) is that of a 6-18-19 near-Pythagorean triangle and, since east

---

*This is 6.83 feet, which is like Scottish engineer Alexander Thom's megalithic rod of 6.8 feet, proposed by him at Carnac. The fact that 27.32166 can be resolved to a near integer within three lunar orbits could be exploited to make the measures every 82 days and record them as a set of stones across the alignment, with each set 82 inches from the previous stone's megalith. (see discussion surrounding fig. 3.32 in Heath, 2014.)

†The Metonic period of 5 × 47 lunations is actually just over 20 (5 × 4) eclipse years long, making it also a weak eclipse cycle, significant instead as embracing the whole ensemble of solar lunar variation.

and west are accurately maintained at both the the Crucuno and Kerzerho rectangles, then the 18 side might have been assumed (for information on Crucuno see chapter 1). Today, there is no monument at the right angle of the triangle, but one can see, in figure 2.14, that the 18 side was arced down to the area where the alignments terminate on the east, probably to avoid the observatory hill of Mané Braz while also providing a counting corridor for the Saros period of just over 18 years.

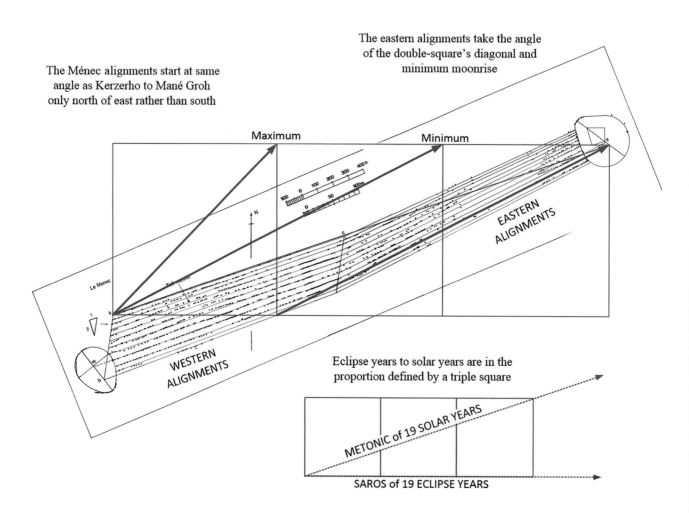

FIGURE 2.13. The overall form of the Le Ménec alignments exploits near identical moonrise locations that occur every three orbits, or 82 days apart.

Survey by Alexander Thom, 1972, published as Thom and Thom,
Megalithic Remains in Britain and Brittany, *1978.*

Locating monuments more precisely using Google Earth

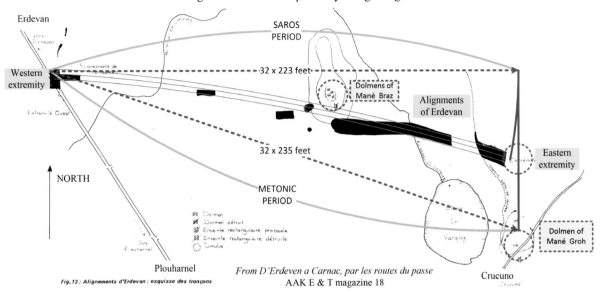

FIGURE 2.14. Alignments near Erdevan based on counting lunar months of the Saros and Metonic Periods.

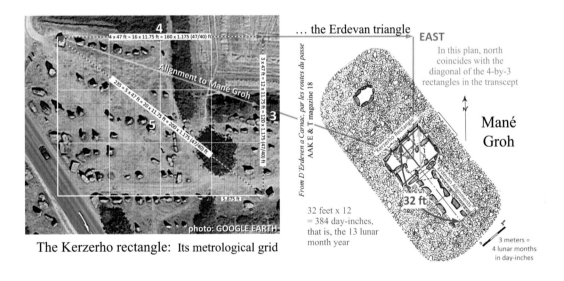

The Kerzerho rectangle: Its metrological grid

… the Erdevan triangle

EAST

In this plan, north coincides with the diagonal of the 4-by-3 rectangles in the transcept

Mané Groh

32 feet x 12 = 384 day-inches, that is, the 13 lunar month year

3 meters = 4 lunar months in day-inches

FIGURE 2.15. The tumulus of Mané Groh contains a model of the stones at Kerzerho, the beginning of the triangle of fig. 2.14. The tumulus is a tilted model of the alignments starting at Kerherzo (*left*) founded on a {3 4} rectangle and Mané Groh (*right*) is found along the alignment of the Third Triangle. Its "transcept" represented Kerherzo while its corridor represented the Erdevan alignments.

Image underlay for figures 2.14 and 2.15 from Association Archeologique Kergal's 1980s magazine Études et Travaux, *vol. 7, p. 54. See also Heath (2014), 62–63.*

The Type B "flattened circle" geometry, of third-millennium Britain, can be seen in the context of the early fourth-millennium Le Manio Quadrilateral, near Carnac (fig. 2.16). The latter's four-square geometry, over 4 lunar years (from point *N*), can be truncated to three-squares (from point *P*—the solstitial "sun gate"), generating the Third Triangle for 3 solar years on its diagonal (to point *J*) and 3 eclipse years, on the base between stones 1 and 34 (see bottom of fig. 2.16). An even earlier, mid-fifth-millennium version of this triangle

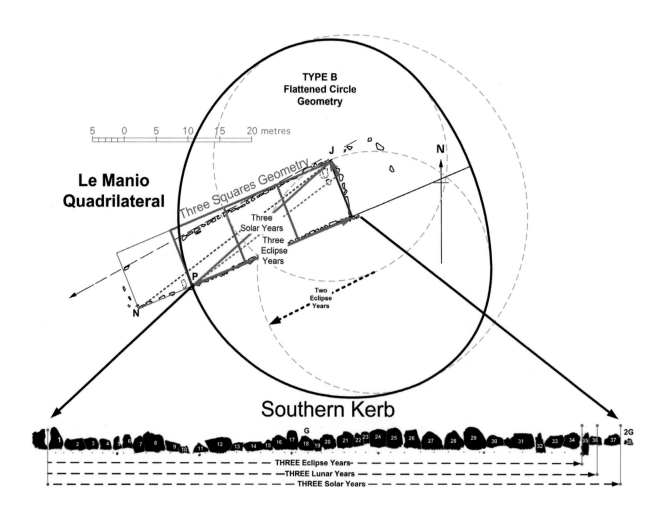

FIGURE 2.16. The Le Manio Quadrilateral (*above*) in the context of a Type B circle also shows three squares as a possible subset of four squares (points *N* to *J*); the Southern Kerb stones in silhouette (*below*) show stone 35 as representing an eclipse year from point *P*, from stone 1 to stone 34.

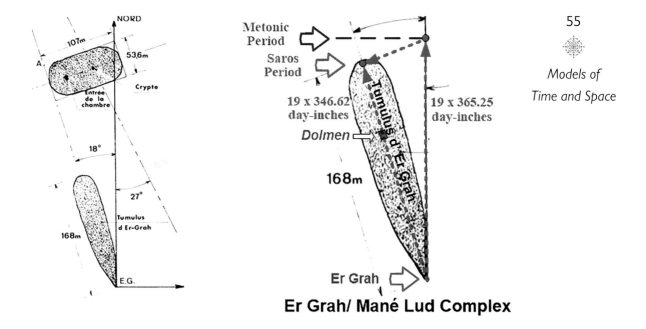

FIGURE 2.17. The length of tumulus Mané Lud (*left*) fixed the triangular layout of Locmariaquer between its northeastern extent to the grand menhir so as to define (*right*) the angular alignment relative to north of tumulus Er Grah.

Image underlay from Association Archeologique Kergal's 1980s magazine
Études et Travaux, *vol. 7, p. 54. See also Heath (2014), 62–63.*

can be found at Locmariaquer to the east, plainly giving the two sides of the near-Pythagorean triangle (in day-inches) for 19 solar years and 19 eclipse years (fig. 2.17).

The eclipse year is 18.618 days squared (346.62 day-inches), and the monument demonstrates a fluidity of concept toward the node day as 18.618 days for the nodes to move a day in angle: 19 years of 365.2422 days equals 19 × 19.618 × 18.618 node days, while 19 × 346.62 days is 19 × 18.618 × 18.618 days. The numbers 18.618 and 19.618 have been factored into the world of time through the orbit of the Moon's nodes, which are numerically explicit in the Moon's nodal cycle relative to the solar year (the orbit of the Earth) and the solar day (its rotation) so as to create the *normalized* form of the Third Triangle—which can be no accident of nature.

This ratio is then expressed in the implied geometry of equal areas. For example, in the first phase of Stonehenge, the 56-hole (7 × 8) Aubrey circle has a radius in day-inches that is one-quarter of the nodal cycle of 6800 days. Figure 2.18 shows a circle of radius 1700 day-inches, which, divided by 18.618, gives a unit equal to one-quarter of a solar year,* so that a square of side 33 is 8.25 solar years or 3013.25 day-inches.

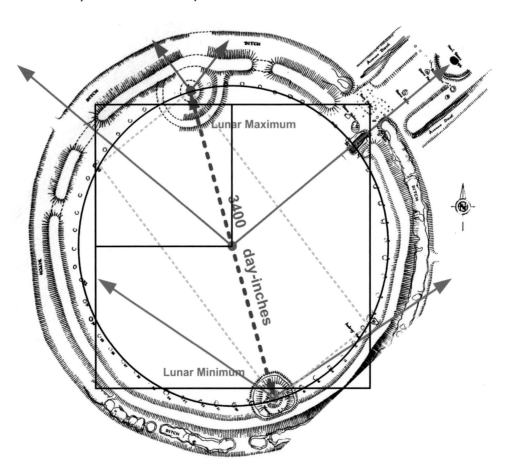

FIGURE 2.18. The early phase of Stonehenge appears to count the 3400 day-inches of half the Moon's nodal period between the diametric Station Stones, whose rectangular nature was known to align the structure's longer sides to the lunar maximum standstills on the horizon.

From Heath, 2014, fig. 6.5, taken from original survey plan from 1972, courtesy A. S. Thom, published as Thom and Thom, Megalithic Remains in Britain and Brittany, 1978.

---

*The model at Stonehenge is on a quarter scale since the radius is one-quarter the nodal period.

The earlier section "The Squaring of the Circle in Area" showed that 18.618 is the smallest number that enables an integer number of years (33) in the side length of a square of equal area to the circle, when that is the radius in years. The solution requires more numerical coincidences, especially that the value from the eclipse year to the solar year, times 19, be almost identical to 18.618, that is, 19.618 (see p. 52, fig. 2.13, *right*). For this to happen, the solar year must be 19.618 × 18.618 days long, while the eclipse year must be 18.618 × 18.618 days long. It therefore seems that 19 was chosen to be equal to these factors—of node days (moving one DAY in angular motion in 18.618 days of solar motion upon the Ecliptic)—which is achieved through recognizing that 18.618 and 19.618 in the triangle most strongly relate to being eclipse years of 18.618^2 days.

Figure 2.19 reveals that right triangles that normalize to superparticular numbers with a fractional part of 0.618 deliver an intermediate hypotenuse that then "lands" as an integer. The outer context might have appeared incomprehensible to an integer-hugging megalithic astronomer, but the inner coincidence causes the eclipse and solar years to have the coincidence over 19 years called the Saros and Metonic periods, whose durations are separated

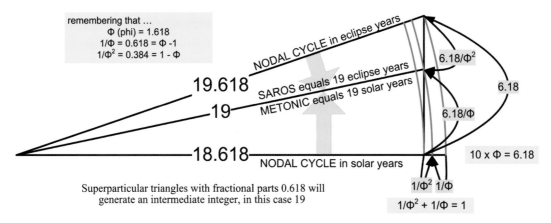

Angelic Geometry deriving SAROS and METONIC from the NODAL CYCLE
as **19** eclipse and solar years, respectively

FIGURE 2.19. The Golden Mean point on the third side enables
18.618 and 19.618 to give birth to the integer 19.

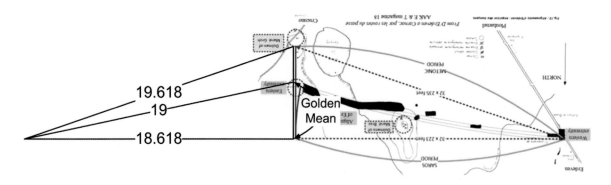

FIGURE 2.20. The existing Erdevan alignments (seen here as fig. 2.14 flipped upsidedown) are revealed as terminating at the Golden Mean point of the Third Triangle's third side, which defines the integer 19.

Image underlay from Association Archeologique Kergal's 1980s magazine Études et Travaux, *vol. 7, p. 54. See also Heath (2014), 62–63.*

by exactly 1 lunar year, since 235 minus 223 equals 12 lunar months.

The next chapter will examine what is being shown by the means of geometrical models, within the domes and other monuments that are examined in the chapters after that, revealing a continuity of knowledge of something called many names in different myths worldwide, such as Hamlet's Mill, Amlodhi's Quern, and the Churning of the Ocean, from a regionally diverse, yet congruent, tradition of cosmological myths pointing to Great Time.

3

Measurements of the Earth

ALONGSIDE ASTRONOMICAL TIME-MODELING there were two models of the Earth, the second only implicit within ancient metrology and for which explicit evidence, of the surveying necessary to have achieved it, is harder to find. The simpler Equal Perimeter model may be megalithic, but third-millennium Egypt and the Classical period Parthenon and Pantheon evidence a model close to the modern levels of accuracy—in contradiction to modern historical expectations.

THE EQUAL PERIMETER MODEL

The modern mean Earth radius of 3958.775 miles rounds up to 3960 miles, and if this is taken as the radius of the inner circle of the Equal Perimeter Model of 5.5 units, then dividing 3960 by 5.5 gives 720 miles as the length of each unit in the geometry. The mean radius of this model is therefore 5.5 × 720 miles, and the Moon's radius of 720 × 1.5 units, equaling 1080 miles, compares well with the modern figure of 1080.067 miles.[1]

Michell commended to me his little-appreciated treatment of factorials, in his appendix to *At the Centre of the World*, as a very profound mystery, and this unit of 720 will be found in the harmonic model of chapter 5 as a key to musical harmony and tuning, since 720 is factorial 6 or 6! = 1 × 2 × 3 × 4 × 5 × 6 = 720. Factorial 6 provides the three prime numbers {2 3 5}, in the correct proportions for five different musical scales to appear in this number, 720. The matter is illustrated in figure 3.1.

The mean Earth radius, which is 3960 miles, can appear in any units or

1!	2!	3!	4!	5!	6!	7!
1	2	6	24	120	720	5040
					11 × 6!	11! - 7!
					3960	720

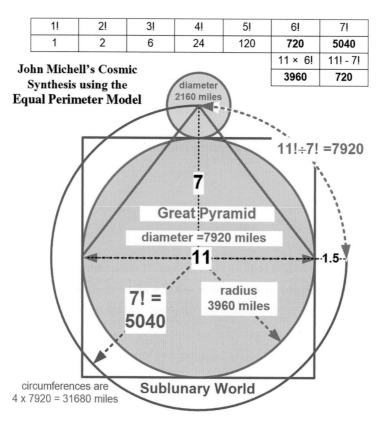

John Michell's Cosmic Synthesis using the Equal Perimeter Model

diameter 2160 miles

11!÷7! =7920

7

Great Pyramid

diameter =7920 miles

11 ---1.5--

7! = 5040

radius 3960 miles

circumferences are 4 x 7920 = 31680 miles

Sublunary World

FIGURE 3.1. Cosmological model of the Earth and the Moon, based on the Equal Perimeter Model.

scale that *preserves the number 396,* including 39.6 feet (e.g., Stonehenge and St. Mary's Chapel, Glastonbury) and 396 inches (Kaaba).[2]

THE METROLOGICAL MODEL

Ancient metrology by historical times used to find and define the size of the Earth more accurately than the equal perimeter approximation ever could. That is, metrology moved from an astronomical focus to developing an accurate model for the Earth. The Equal Perimeter Model of the Earth was probably an early estimate, giving the mean radius as 3960 miles. In actual fact, the mean Earth radius is 3958.775 English miles, and close to 7×12^6 English feet long—to an accuracy of 0.002 percent. This mean Earth radius (R_{mean}) must have been established in ancient times as being 3958.691 miles due to the attraction of it being 7×12^6 English feet long and so, on that basis (assuming π to be 22/7), the mean circumference of the Earth was known as 44×12^6 (using $2 \times \pi \times R_{mean}$), since the mean Earth is, by definition, spherical, and its cross section is a circle.

While the system of ancient metrology is not our main focus, its main modules will be described, some in detail, in appendix 1, and a bibliography provided. We will focus here on the microvariations of the English foot, the root of the whole system. These variations were applied to every major variation of the English foot, as required, to provide numerate solutions in the monuments built using the system. These microvariations were primarily introduced to solve two problems: (a) the maintaining of integers, within circles, between the lengths of their radius and their circumference (176/175), and (b) regarding the difference between the mean and polar radii of the Earth (441/440).

The historically verifiable *types* of foot, called *modules*, are named after the place they were first found: (1) in common use, (2) in local monuments, (3) defined as measures within foot-shaped depressions, (4) in motifs, (5) in reliefs, (6) in images of measuring rods, (7) on the rods themselves, or (8) on other carved geometries. As already stated, sometime in the 1990s, John Neal observed that the root feet in each module were related to the English foot as a simple rational fraction of the foot still in widespread use. Since many historical feet were close to being a simple ratio of the foot, these had a more complex relationship to the English foot due to microvariation, which varied the length of feet within a module relative to the root foot, but for what purpose?

THE SYSTEM OF ROOT VALUE MODULES

Works on historical metrology are confusingly complex despite giving many clues to underlying rational relationships between historical measures and the English foot and, hence, among the historical modules.[3] These systematic applications of the microvariations were invisible to modern specialists; they did their calculations using arithmetic and trigonometry rather than through ratios, not knowing that metrology was a form of pre-arithmetic numeracy based upon ratios.

This variation of measures within each given module had prevented Neal's predecessors from seeing that a missing root value interconnected the system of feet to a super root value of one English foot. The root value of each module, itself a simple ratio to the English foot, placed all the historical measures as orderly expressions of a single common system of metrology found throughout the Old and New World civilizations.

The root value of the Roman module, based upon its historical values, had to be 24/25 feet—a foot rarely found in the field. When *the same* variations were applied to the root (to solve specific types of ancient calculation) as those found in other modules, it was always the same pattern, or grid of microvariations, that appears to have been used.

The Recovery of Metrology

John Michell's work on ancient metrology emerged in 1981, and he described two feet in each named module of historical metrology, which were 176/175 feet apart, identified as "tropical" (southern) and "northern" versions, because he already knew that the north-south degree lengths of latitude 51 to 52 degrees north and 10 to 11 degrees north differed by this ratio, with degrees (or parallels) of latitude gradually growing in length to the north.

For example, Michell's table had the tropical Roman foot as 0.96768 feet and its northern value as 0.9732096[*], while the root, according to Neal, should be 0.96 (24/25) feet.[4] Michell's tropical Roman foot was 1.008 × 24/25, while his Greek tropical value was 1.008 (126/125) feet. Though Michell's northern feet were always 176/175 larger than his tropical feet, his southern feet turned out to be differently offset from their true root value.

Michell's tropical Royal foot was 441/440 of its rational root of 8/7 feet, his tropical Greek foot was 126/125 of its root value (which was actually the English foot), as also was his tropical Roman foot. This indicated to Neal that just two core microvariations existed in ancient metrology, namely:

1. 176/175 = 1.0057—the tropical-to-northern ratio, and
2. 441/440 = 1.0023—the ratio of the mean radius of the Earth to its polar radius

---

[*]It may seem silly to have so many decimal places in the value of a foot length but these fractional values result from using base-10 notation and, over long itinerary lengths, miles long, one can discriminate the different feet to high accuracy.

Their sum, 176/175 × 441/440, equals 126/125 (1.008), the offset of Michell's tropical Roman and Greek feet from their respective root values, while 441/440 was the offset of the tropical Royal foot from its root value of 8/7 feet.

This revealed two main directions of microvariation, which can be organized as a table or grid in which direction to the right was the tropical-to-northern ratio (176/175) while movement at right angles carried the signature of the polar-to-mean radius (441/440). Neal started to work on known variations of feet belonging to different modules and found the majority had a home within his grids. The English module, based on the number 1 for the system, can be used to illustrate this (table 3.1).

TABLE 3.1
MICROVARIATIONS FOR THE GREEK MODULE.*

Operation			× 176/175	× 176/175
	English	**"Root"**	**"Canonical"**	**Geographical**
× 441/440	**"Standard"**	1.00227	1.008	1.01376
	"Root"	1 foot	1.00571428	1.0852905

*This table has 441/440 increasing upward rather than downward as found in Neal's grids. This is purely pedagogical in avoiding the counterintuitive for the sake of the reader new to metrology yet familiar with reading XY graphs. Neal's naming of microvariations will and should be retained, to correspond to his work. The words in quotes are John Neal's definitional vocabulary such that "standard canonical Greek" (1.008) locates a given microvariation called the "Olympic foot" in the prior historical metrology as per Berriman (1953).

A number of these feet are identified within historical metrology as Greek, and so the English foot was revealed, by Neal, to be the root measure of the Greek module. While all historical modules resolve in this way into a root measure, the root value may not be a known historical measure, if it has not been found or identified. The many modules evidently existed because each had a demonstrable and specific utility. This is especially true of measures having the primes 7 and 11 in their formula where, obviously, the utility of π as 22/7 can apply to circles and models of the Earth, as with the Great Pyramid, where the height is 7 units to the southern base of 11 units. The Great Pyramid's height was 440 Sumerian feet of 12/11 feet, while its southern base was 440 standard Egyptian cubits of 12/7 × 441/440; that is, both are the same number

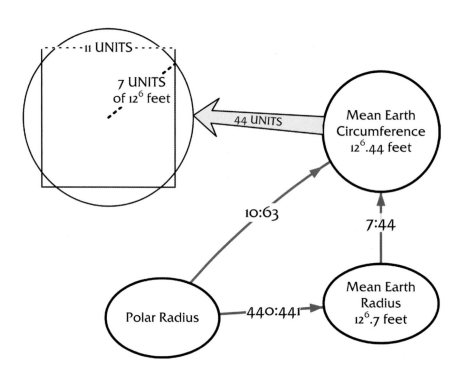

FIGURE 3.2. The ancient model of the Earth based on 12^6 feet,
two approximations to π, and the 441/440 microvariation in ancient
metrology reflecting the ratio of the mean earth radius to the polar radius.

of different units, one employing 11 in the submultiple and the other having a
factor of 7/11. The whole of the Giza complex employed this standard Egyptian
cubit (1.7$\underline{18}$ feet), while 440 Sumerian feet were used in the truncated height of
the Great Pyramid to represent the polar radius, relative to the 441 of its ideal
height, as the mean radius of the Earth. The missing apex (or pyramidion) provided the extra Sumerian foot. You can therefore directly see the ratio 441/440
both in the Great Pyramid and in the standard Royal cubit throughout the
Giza complex.

Returning to the model of the Earth implicit in metrology, it is already
clear that one of the microvariations is the ratio between the polar and mean
Earth radii. The mean Earth radius is 7 × 12^6 feet. With π as 22/7, this makes
the mean circumference 44 × 12^6 feet. Figure 3.2 shows the equal perimeter
circumference and square of side 11, but in this case the mean Earth is not the
in-circle of 11 units but rather the radius of 7 units. This model of the Earth

can be seen in the Pantheon (see p. 68), which uses the method of dividing the diameter of a circular building by 14 to discover its units according to this model of the Earth.

A further feature of the model emerges when the 440 length of the polar radius is related to the mean Earth circumference, in that the circumference of 44/7 multiplied by 441/440 leaves 63/10 as another approximation of 2 × π = 63/10 (6.3). The only absentee is the microvariation 176/175, but this is the required proximate relation needed to maintain integer relations between a circle's radius and its circumference, for 176/175 = 4/25 × 44/7. For example, if a circle's radius is 12 feet, its circumference using a π of 22/7 is 75.4̲2̲8̲5̲7̲1̲ (528/7) feet but, using a foot larger by 176/175,* the circumference resolves to 75 feet: the larger foot, divided into 75.4̲2̲8̲5̲7̲1̲, has *removed the fractional part* because 176/175 = 16/25 × 7/11, counteracting 2π as 44/7.

The other microvariation is 441/440 and, it too is a proximate composite of 63/10 × 7/44, and so three different approximations to 2π were built into the system: the accurate one of 44/7 and two inaccurate ones of 25/8 and 63/10, one for each of the microvariations.

Backing off a little from metrology, one can see the workings in this of an angelic solution that chose to make the oblateness of the sphere of the Earth such that the mean radius of its ideal spherical shape would be 441/440 of its polar radius. Due to this, various latitudes became rationally related, in their north-south length, to other latitudes, as Neal showed in 2000 in his book *All Done with Mirrors* (see fig. 3.3, p. 66).

The equator is a length that, divided by the number of days in a solar year, gives 360,000 English feet so that 1 DAY on the equator defines a length similar to the increasing lengths of degrees as one travels north. Divided by 360,000, each of the equatorial lengths yields what is called the latitudinal foot,† and what is obvious is that through the microvariations one can know the lengths of at least five latitudes and their latitudinal feet (see table 3.2, p. 66).

---

*Sometimes called "larger by one 175th part" just as 441/440 is "larger by one 440th part."

†For example, the reference length of the parallel of latitude for Stonehenge from 51 to 52 degrees is 364,953.6 feet which, divided by 360,000, gives 1.01376 (3168/3125)—called the "standard geographical" microvariation achieved through (176/175)² × 441/440.

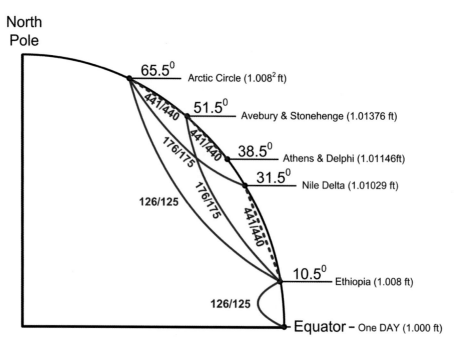

FIGURE 3.3. The interval ratios between latitudinal foot lengths
on the meridian.

Adapted from John Neal, All Done with Mirrors, *2000.*

TABLE 3.2

PLACES ON LATITUDES WHOSE FEET
ARE LINKED BY THE MICROVARIATIONS

"Root"	"Canonical" = x 176/175	"Geographical" = x (176/175)^2
I foot		Athens and Delphi (1.01146 feet)
x 441/440	Ethiopia (1.008 feet)	Avebury and Stonehenge (1.01376 feet)
x(441/440)^2	Nile Delta (1.01029 feet)	Arctic Iceland (1.008^2 = 1.016064 feet)

The Ethiopia-to-Stonehenge ratio is 176/175, which is Michell's tropical-to-northern discriminator. This is a property of the oblate spheroid in which the volume is conserved from the sphere, and the polar radius is reduced by one 441st part. This act appears to have created the form of ancient metrology and its extraordinary ability to provide an Earth that was intelligible through numbers.

One retrospective argument *against* ancient metrology is that if all these variations are assumed, any length found from the ancient world can be shoehorned into the service of a fantasist, to give lengths meanings that were never intended by the builders of something that length. However, after sufficient experience in working with this system, very unusual facts emerge, known to "the ancients," and verified in detail as, in fact, invariant properties of the number field and of the Earth itself. My explanatory framework of angelic action is reasonable since physical laws, as far as we are aware, cannot create higher levels of order in this way.

These variations are then seen in the variation of latitudinal feet as successive parallels increase their surface lengths as latitude increases, as in figure 3.4. A latitudinal foot is the length of the degree in feet divided by 360,000 feet. At Avebury, for example, the degree length is 364,953.6 feet and the natural foot for that latitude, dividing that length by 360,000, is 1.01376 (3168/3125) feet—the "standard geographical."

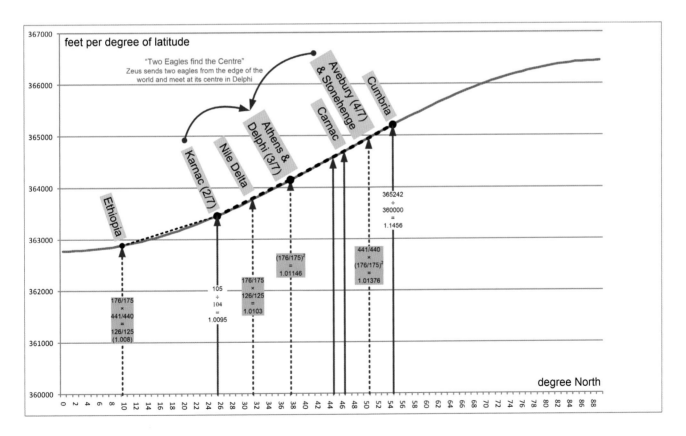

FIGURE 3.4. The increase in degree length with latitude.

Avebury is within the degree of latitude 51–52 degrees north, which has the same length on our nonspherical Earth as all the degrees of latitude *would have*, on the spherical mean Earth. The location of Stonehenge therefore references the mean Earth, and John Michell's transition, from the Equal Perimeter Model (radius 3960 miles) to this metrological model (radius 7×12^6 feet) has maintained the importance of sacred boundaries being numerically similar to 3168, through the mean Earth foot (of this 52nd parallel) being 3168/3125 feet. Between Stonehenge and Avebury, Michell found exactly one-quarter of a degree of latitude, implying Stonehenge was built one-quarter of a degree north of Avebury after an ancient geodetic survey (c. 3000 BCE) had been concluded—establishing its latitude as four-sevenths of the meridian from the equator to the North Pole. By studying the variations built into metrology, one sees the megalithic viewpoint that the shape and size of the Earth was a design based on the number field. Through this, the angelic world had transmitted its primary geometrical models and methods that became the basis of our own ancient geodetic model. This could even be a design for planets with life in general. Quite naturally, sacredness became connected to models of the Earth.

MODELS OF THE WORLD

Michell noted, "The lengths of the various units of measurement have never been precisely defined, and one of the objects of this essay is to provide those definitions as a means to establishing the dimensions of the earth, as formerly reckoned, and of the monuments, notably Stonehenge, which were designed as small-scale imitations of it."[5]

The Pantheon Model

Surprisingly, the Romans built a metrological model of the Earth, the Pantheon, built by the Roman emperor Hadrian and dedicated by 126 CE. It remains the largest unreinforced cantilever made of concrete. The concrete came from the volcanic rocks on which Rome sits. The Pantheon's rectangular portico leads to a large rotunda whose coffered concrete dome ceiling has a central opening, called the *oculus,* or "eye," to the zenith. Pantheon means "of, relating to, or common to all the gods," and the idea of a portico leading to rotunda was revived in recent centuries as the neoclassical style.

John Neal's measurements of the tiled floor confirmed English mathematician and astronomer John Greaves's 1639 result as being within the Roman module: called the Cossutian Roman foot (0.96768 feet), in Neal's lingo the standard canonical Roman (times 126/125 = 1.008) of the Roman foot whose root value is .96 (24/25) feet.

One needs to know the diameter of the inner dome:

1. Wikipedia has the diameter as 43.3 meters (142.06036 feet).
2. A plan (by A. Desgodetz) with a scale gives 43.256 meters (141.929 feet).
3. Neal (working in feet) has 142.24896 feet (43.35748 meters), and the historical figure of approximately 147 Roman feet, of the floor, is the diameter Neal used of exactly 147 of the standard canonical Roman feet (0.96768 feet) as found within the floor: thence Neal's datum.

But Neal then used the 14-unit diameter found in the model of the Earth to divide the dome's 147 *Roman* foot diameter by 14, giving a unit of 10 *Greek* feet, each 1.016064 feet, or 1.008 squared.*

This Greek unit (of 1.016064 feet) was used as a method to transform the 440-foot circumference of the implied sphere held within the Pantheon into 441 geographical Greek feet, which are one part in 441 smaller, that is 440/441 of the foot identified within the *perticae*. Figure 3.5 shows how the sphere would fit between the dome and the floor.

The rotunda is decorated with five rows of twenty-eight "coffers"[†] whose size reduces with the closure of the rotunda, which then has a disc of plain cement with a circular window at its center, the oculus (fig. 3.6). The rotunda can be

---

*Neal suggested that we apply a rule of thumb to circular structures, in this case that the diameter of 142.24896 feet be divided by 14 to reveal the module used, or some multiple of it (Neal 2017, 26–28). The unit he found in this way at the Pantheon was then 10.16064 feet, or a *pertica* of 10 feet, whose feet are the square of 126/125, the standard canonical English foot. He then calls 1.016064 feet "a natural extension of core values," being "the 136,000th part of a meridian degree of latitude in Iceland, the most northerly degree (just below the Arctic Circle)," which will remain meaningful to interpreting the dome's coffers. However, I believe the squaring of 1.008 was for another very elegant reason, suggested by the plan of the Pantheon seen in figure 3.5 (top), which shows that a sphere fitting the dome's interior rotunda would also touch the floor of the building.
†A *coffer* (or *coffering*) in architecture is a series of sunken panels in the shape of a square, rectangle, or octagon in a ceiling, soffit, or vault.

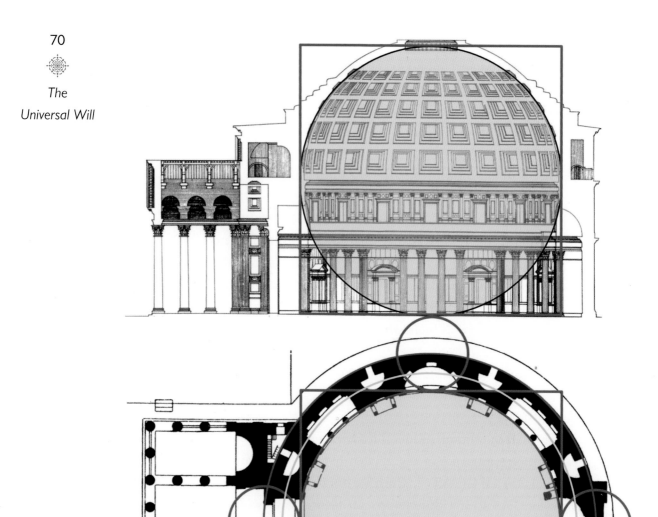

FIGURE 3.5. Cross section of the Pantheon, showing how a sphere of that diameter fits under its dome (*top*). Plan view of the Pantheon, showing its conformance to the model of equal perimeter (*bottom*).

FIGURE 3.6. The Pantheon ceiling from below.

Photo by Mohammad Reza Domiri Ganji for Wikimedia Commons.

interpreted as representing the seven sacred latitudes of the ancient world that approach the northern latitudes, up to the Arctic Circle (the plain disk): the five latitudinal zones, then the Arctic disc, and then the North Polar eye.

Assuming 360 degrees in a circle, 360/28 = 90/7 = latitude 12.<u>857142</u> degrees, which multiplied by 5 equals 64.<u>285714</u> degrees at Iceland.

TABLE 3.3

THE FIRST FIVE SEPTENARY LATITUDES

Location by Latitude North	Corresponding Place
12.<u>857142</u>°	Ethiopia
25.<u>714285</u>°	Karnak
38.<u>571428</u>°	Delphi
51.<u>428571</u>°	Avebury
64.<u>285714</u>°	Iceland

Most obviously, then, the inner space is a model of the world from floor to oculus, while the dome is a model of the Northern Hemisphere. The Equal Perimeter Model gives a first approximation to the shape and dimensions of the Earth and the Moon, but, as we have seen, ancient metrology modeled the actual size and shape of the Earth using the ratio 176/175 as well as 441/440, the mean-to-polar radius found in both models. The later model refined the mean Earth radius from 3960 miles to being 3958.6$\underline{90}$ miles, which is 7×12^6 feet (see fig. 3.2, p. 64). The 7 in the radius of the mean Earth means the spherical mean Earth circumference is 44×12^6. The 12^6 foot unit that is in common can be thought of as the common units represented by the perticae of 10×1.008^2 in the Pantheon. Substituting 1 for 12^6, the diameter equals 14 and the circumference equals 44, so that when 1 becomes a 10-foot unit, the diameter becomes 140 Greek feet and the mean circumference 440 Greek feet, and these base numbers define the spherical Earth held between the floor and the domed ceiling of the Pantheon.

When the 10-foot unit is 10×1.008^2 feet, which are 441/440 of the geographical foot of 1.01376 feet, the circumference is also 441 of perticae in feet with the geographic foot ratio of 3168/3125—1.01376 feet.

This fulfills John Michell's discovered rule for sacred perimeters: that they followed a rule of measuring, in ancient units, 3168 units or some simple variation such as 316.8 feet to then represent the mean Earth as a spiritual ideal.[6] The geographical Greek foot contains such a factor: when the mile of 5000 feet becomes $5000 \times 3168/3125 = 3168 \times 8/5$ feet, these then divide the mean circumference as 25,920 geographical Greek miles.

Only a measure 1/440 greater than the geographical foot could, times 10, provide the 440 division of the Pantheon's great circle, which then equals 441 geographical Greek feet. The Pantheon's form and size emerged exactly out of these requirements in that the Pantheon is the smallest spherical space in which the geographical Greek foot could express the 440 units of the model of the Earth's polar radius while expressing the 441 units of that model's mean radius of the Earth. Why is a matter for contemplation. But there can be no doubt the Pantheon was a profound demonstration of the metrological model of the Earth, and, as we shall see in chapter 4, domes in particular seem to have their origins as being models for the spherical Earth.

4

Temples of the Earth

THE NUMERICAL RELATIONSHIPS found in ancient models of the Earth and its surrounding Moon, Sun, and planets are the basis of many sacred buildings. These relationships are reimagined on the Earth to form sacred spaces that are then numerically related to the higher levels of order that are the true prototypes, archetypes, or models of sacredness. By building structures on our planet that relate to the cosmos, the form of the heavens was brought to the Earth in a long-established manner that introduced humankind (at least subconsciously) to places with a special energy. Ecclesiastical institutions place the sacred constructions within their own jurisdiction, but there is an uneasy truce between sacredness for its own sake and the need to control a religious franchise.

The earliest buildings designed to reflect the cosmos in this way were probably not sacred as such but built for the purposes of astronomical discovery and the quantification of angles and lengths (that is, their purpose was *to realize*). Since then, reproducing the cosmic interrelationships had to be through representation of the sacred geometries, numbers, and ratios. The Lunation Triangle, for example, could be couched within the Second Triangle or seen as a four-square rectangle, representing the discoveries made. Sacredness is not generally triangular but rather uses the square and circle to enclose a triangle with numerical relationships belonging to the sacred models of the equal perimeter or equal area, which provides the main theme of this chapter.

Both models are to be found in the Great Pyramid, a development from stepped pyramids in Egypt. In Babylonia, the ziggurats had terraces that appear, like the Pantheon, to model the key latitudes of the Earth's Northern Hemisphere. The Great Pyramid mapped, in the mathematical sense, the four quadrants of the Northern Hemisphere onto each of its triangular faces, while also integrating the models of equal perimeter and equal area. It did this through integrating two of the most significant cosmic irrationals, π and *phi*, maintaining whole-number transforms by manipulating whole-number approximations of irrationals, such as *phi* by the Fibonacci series. The section after next discusses how pyramids and Greek temples took the form of domes based on the equal perimeter and area models. But first, let's discuss pyramids.

Great Pyramid of Giza, 2560 BCE

Piazzi Smyth, nineteenth-century Astronomer Royal for Scotland, linked a length for the King's Chamber, within the Great Pyramid, to the base-side length of

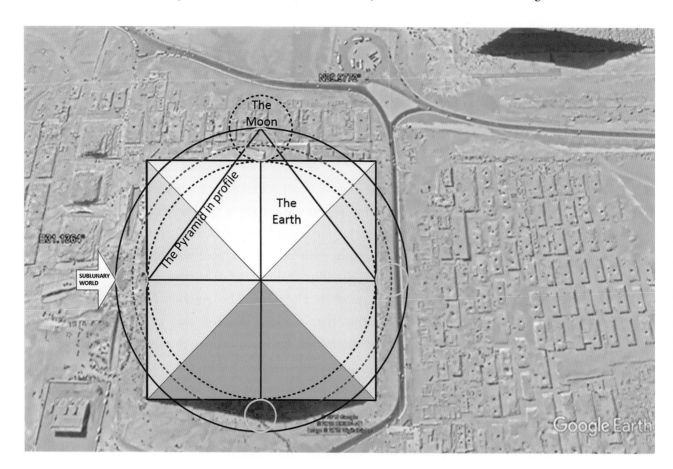

Figure 4.1. The Great Pyramid environs with both area and perimeter models shown.

that pyramid. He showed that, to reasonable accuracy, the *number* of inches in its length could be seen as *the same number* of sacred cubits relative to the southern side length of the pyramid's base, so as to form the Equal Area Model (see fig. 4.2, p. 77, adapted from *The Great Pyramid*, 1880). We have already seen that this pyramid has the 7 high and 11 across profile that stands at the heart of the Equal Perimeter Model. Thus, in 1880, Smyth interpreted the square base in his diagram called "Equation of Boundaries and Areas." The pyramid's shape, while geodetically profound, may also have had meanings relating to the ratio of 18.618 to 33 (the square root of π) based upon the area of its base, making that base a visual model of the nodal period relative to 33 years, the number of the solar hero when the rising Sun recurs on the exact same spot on the horizon.

From this combined model emerges some understanding of a notion, found in Plato and Aristotle, of the *sublunary world*. Wikipedia defines this as:

> The sublunary sphere was the realm of changing nature. Beginning with the Moon, up to the limits of the universe, everything (to classical astronomy) was permanent, regular and unchanging—the region of aether where the planets and stars are located. Only in the sublunary sphere did the powers of physics hold sway.[1]

This has been interpreted in modern times to say that the laws of physics belonged only to what lay within the lunar orbit, which is certainly not true. It is true though, that many things that *can happen* on the Earth are completely different from those things that happen in the rest of planetary space and, significantly, it seems that life has not emerged in a similar way on the other planets—though that would not be known without modern instruments. There was a clear association in the medieval period—which had inherited this sublunary model—that death was associated with the domain beyond the Moon and our lives within the sublunary sphere.[2]

Throughout the megalithic and ancient worlds, lunar behavior appears associated with alignments to the Moon's extremes, and lunar eclipses evidently had a high profile for counting the moons in between them. But possible references to this in mythology have not been made clear. A modern contribution was from Professor Sir Fred Hoyle (1919–2001) who, in his *On Stonehenge*, speculated that the nodes were some kind of third force for

the Sun and the Moon, rather like the Holy Ghost in the Catholic Trinity.*
In the perimeter model, the sublunary space is empty, but by concentrically
combining it with the area model, the circle of equal area *to the same square*
runs through the middle of the sublunary, showing the grinding nature of
the lunar orbit's retrograde precession via the rotation of its nodes over 18.618
years; a *similar action* to the Precession of the Equinoxes over the much longer
period of about 25,920 years (for its myths see *Hamlet's Mill,* de Santillana
and von Deschend, 1977). Through the area model, the Great Pyramid might
have presented the lesser mill of great time where the Earth is the bottom
stone of the mill and the lunar orbit the rocking stone beneath which the
wheat of existence is ground up to feed the gods and maintain their immor-
tal battle with the demons, a battle described in the myth-made-monument
of the "churning of the ocean" at Angkor Wat. The demons are presented as
essential, if the potentials of the spinning-world churn are to be maintained,
out of which (we are told) wonderful objects appear.

The two forms of "squaring," by area and perimeter, can be combined
through the geodetic insight that *numerically* the equator of the Earth is the
solar year in days multiplied by 360,000 feet. When Smyth wrote *The Great
Pyramid* (1880), he noted (in his fig. 4 of plate XXI) the sensational result that
the side of the pyramid is 365.24 sacred cubits, that it is related to the solar
year in days. The solar year therefore became connected, within metrology, to
the size of the Earth. And since the solar year is 365 + 32/132 days (and 132
is 4 × 33), the 33 years of the area model (the side length of the base in the
area model) could also be seen alongside the mean Earth diameter, linking the
movement of the lunar nodes (that cause eclipses) with the orbit of the Earth,
the Great Pyramid already embodying the Equal Perimeter Model of the Earth
and the Moon (see chapter 2).

Smyth's sacred cubits were double feet of the common Greek foot of
36/35 feet, anciently held sacred to both Egyptian and Hebrew civilization. The
length of the King's Chamber is 20 Gaza Royal feet or 34.<u>36</u> feet and hence

---

*"Was the [nodal entity] a still more powerful god, an unseen god able to overcome the visible
deities? Could this be the origin of the concept of an invisible, all-powerful god? . . . With the later
demise of the Sun god, the god-like quality of [the nodes] would seem to have remained, to become
the Invisible God of Isaiah. Could the distant memory of Sun, Moon and Nodes be the origin of
the doctrine of the Trinity, the "three in one, the one in three"? Hoyle, *On Stonehenge,* 133.

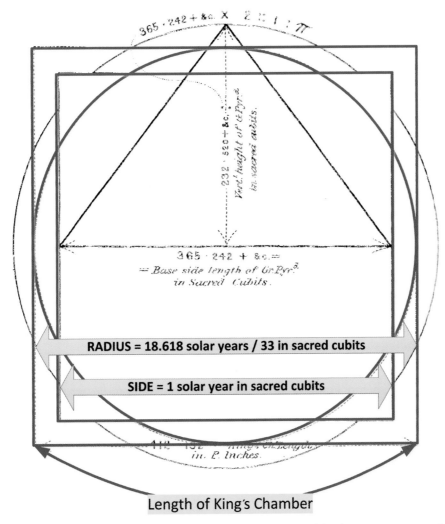

RADIUS = 18.618 solar years / 33 in sacred cubits

SIDE = 1 solar year in sacred cubits

Length of King's Chamber

FIGURE 4.2. This diagram shows the scale used to enable the Great Pyramid to square the King's Chamber, of inches to sacred cubits.

Image underlay from Smyth, The Great Pyramid, *part of plate XXI.*

142 inches (ignoring the virtually identical Pyramid inch of 1001/1000 inches). The area model would divide the pyramid's base by 33 and then multiply the unit obtained by $2 \times 18.618 = 37.236$, the diameter of the circle of equal area. The diameter is then 853 feet, which is 409 sacred cubits, which more or less matches Smyth's 412 inches of the King's Chamber. His assumptions and calculations were different from mine, but his intuition, that inches within the chamber equaled the required diameter in sacred cubits (for the area model to

the base of the pyramid), was substantially correct, and this might have been intentional, since the said chamber length is 33 sacred cubits long!

Once the pyramidology craze had been curtailed by the scientific establishment, the last metrologist of note was Sir William M. Flinders Petrie (1853–1942), who surveyed the Great Pyramid to modern levels of precision in the late nineteenth century.

> In his teenage years, Petrie surveyed British prehistoric monuments . . . in attempts to understand their geometry (at 19 producing the most accurate survey of Stonehenge). His father had corresponded with Piazzi Smyth about his theories of the Great Pyramid and Petrie travelled to Egypt in early 1880 to make an accurate survey of Giza. . . .
>
> Petrie's published reports of his triangulation survey, and his analysis of the architecture of Giza therein, was exemplary in its methodology and accuracy, disproved Smyth's theories and still provides much of the basic data regarding the pyramid plateau to this day.[3]

In 1877, before going to Egypt, Petrie published his *Inductive Metrology*—having realized that, instead of relying on the ancient authors who had given measurements and units, one could and should use the direct means of making measurements, in the field, of surviving monuments and other items such as measuring rods, metrological statues, and so on. However, by 1920, the learned essay in the *Encyclopaedia Britannica* on metrology was left out of subsequent editions. This marked a period of neglect for ancient metrology, brought on perhaps because anyone (like myself) could then measure monuments and create further claims like the Victorian pyramidologists had.

The subject was eventually resurrected by nonacademic research into monuments by John Michell (1982) and John Neal (2000). Valuable compendia such as Berriman's *Historical Metrology* (1952) failed to comprehend the underlying structure of ancient metrology—mixing all manner of weights and measures rather than concentrating on the linear measures within built structures.

When we look at the metrological facts, it is not true that all of Piazzi Smyth's theories were wrong. Smyth did not know to see 33 years and 18.618 as definitional for the area model—though he was imposing the known geometries of the perimeter and area models onto the pyramid's plan. Comparing 33 years (instead of just 1 year in Smyth), the intended south-side length of the pyramid,* we are then comparing 2 × 18.618 years = 37.236 years to Petrie's length of the King's Chamber in inches. Smyth's ratio was 1.12838, while 37.236/33 is 1.1283<u>6,</u> which is effectively the same. This means that the Pyramid inch in Smyth could be seen as scaled up from the King's Chamber length in inches to *the same number* of sacred cubits, which he says were made up of 25 inches and hence were the *standard canonical* sacred cubit. The greater length is 756 feet × 1.128<u>36</u> = 853.0429. The side length of 756 feet is 800 Samian feet of 0.945 feet (as recorded by Herodotus around 454 BCE), while 853 feet is close to 800 Persepolis feet of 16/15 feet (a microvariation of the Samian foot).

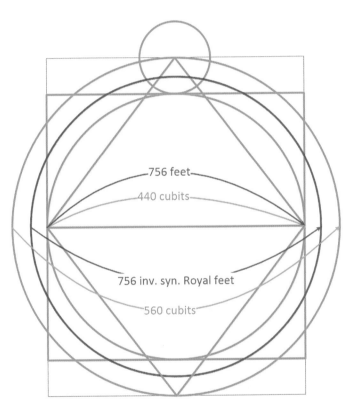

FIGURE 4.3. The 18.618 radius circle is the same as the *number* of feet of the Great Pyramid's side length × 80/81 × 8/7. How this was symbolized within the curtilage remains to be seen.

---

*Neal (2000) has pointed out that the Great Pyramid's sides were deliberately given different lengths, in proportion to four of the parallels of latitude found in Egypt: (east) the Great Pyramid's latitude, (south) that to the north (the Nile delta), (west) that immediately south of the pyramid, and (north) that of Karnak-Thebes, but the reference length is the south side of exactly 756 English feet (after Petrie's defining survey). The idea of averaging all the sides of the base has confounded this fact and may well have led to inaccuracies for Smyth's numerical equations for which, today, one needs to make allowances.

Adopting these measures gives the ratio between them of 1.12875, the ratio between 37.236 and 33, showing that a solution exists within ancient metrology, relating the desired length of 583 feet and the Pyramid's side length. My own analysis combines the area and perimeter model at the Great Pyramid as shown in figure 4.3, and where indicated in other monuments.

The Area Law for the Moon

It seems the numerical relations of the equal area model are due to the Moon being the Earth's satellite. One can then turn to Kepler's resolution of planetary dynamics, in which planets sweep out equal areas in the same time in different parts of their elliptical orbits around the Sun.

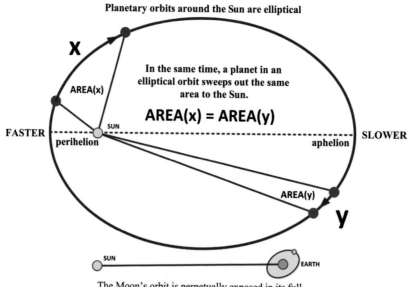

Planetary orbits around the Sun are elliptical

In the same time, a planet in an elliptical orbit sweeps out the same area to the Sun.

AREA(x) = AREA(y)

The Moon's orbit is perpetually exposed in its full area so that, in 18.618 years, its nodes sweep out an equal area to a square of side 33 years, the basis of the Equal Area Model.

FIGURE 4.4. Kepler's second law concerns areas swept out by planets within elliptical orbits as being equal whatever their speed at different parts of their orbits. The Moon's orbit, as a satellite of the Earth, is constantly exposed to the Sun, and thirty-three Earth orbits come to relate as a square area to the nodal period as a circular area, according to π as the Equal Area Model.

This law is obviously different for the lunar orbit around the Earth, and yet the Sun is a powerful gravitational force on the Moon since the orbit is oblique by 5.1 degrees to the solar plane joining the Sun and the Earth. It would therefore appear that the area swept out by the Moon's nodes in 18.618 years resonates with the Earth's orbit and rotation, which takes 33 years to cancel the 0.2422 fractional part of the year *in days of rotation*. Further analysis shows that this is due to cancellations based on the 1/33rd part of the solar year of 11.0679 days, hence giving 33^2, while the excess of the solar year over the eclipse year of 346.62 days gives the 18.618^2 in the $\pi = 33^2$ divided by 18.618^2 (see fig. 2.10, p. 47). In contrast, precession is caused by the angle of the Earth's rotation within the gravitational field of the solar system, on the Ecliptic.

THE PARTHENON MODEL, 400 BCE

The Parthenon was built around 400 BCE on a pediment exactly 101.25 feet in width aligned to true north, which is obviously 100 feet of a foot of 1.0125 or 81/80 feet, making it, like many such Greek temples, a "hundred footer" (*hekatompedon*) in its width. This is an accurate *average* length for 1 second of angle on the Earth's meridian.

This would indicate a replacement, by Classical times, of the geographical constant of 1.01376 feet within the model of the Earth since the original model, by the late megalithic period, had assumed that the meridian was exactly half

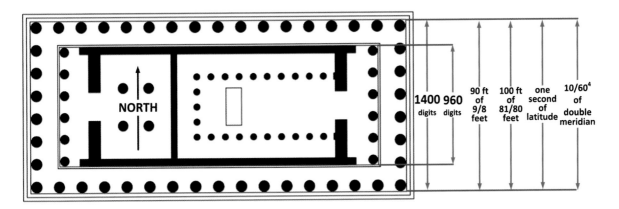

FIGURE 4.5. The geodetic significance of the Parthenon's pediment.

of the mean circumference of the Earth. These alternative geographical constants coincidentally represent the ubiquitous theme in ancient musicology, of the transition from ancient Near-Eastern musical theories and the Pythagorean and Just octave scales and their respective commas called Pythagorean and synodic, respectively.

An article by Jay Kappraff and Ernest McClain observed that the width of the Parthenon's pedement (or platform) symbolically defined 1 second of latitude within 0.003 percent (taking surface lengths as linear fractions of latitude) of its modern estimation.[4] This implies that its intended north-south design was smaller than it would be had the circumference of the mean Earth been referenced as was normal for monuments built two thousand years earlier, such as Stonehenge and the Great Pyramid of Giza. The actual meridian length of the Earth is due to what is called its geoid or nonspherical shape.

In the ancient model of the Earth, three different approximations of π were used to allow for the distortion of the rotating planet over what its mean, or perfectly spherical, circumference would be. In that model, the circumference of the mean Earth was taken as being identical to the actual, double meridian length, both lengths then assumed to be 44×12^6 (131,383,296) feet, or 24,883.2 miles[5] (see chapter 3, "Models of the World"). Were the Parthenon to have been built using this older but sophisticated model, then its shorter sides would be 101.376 feet in length, and one-hundredth of this would be a foot of 1.01376 feet, the foot known as the standard geographical Greek foot. (This fraction of 1.01376 [3168/3125] is called the geographical constant* because it relates to the variation of north-south degree lengths of latitude due to the shape of the Earth† and defines the meridian as half the length of the mean Earth's circumference.)

The double meridian in the model (equal to the mean circumference of the Earth) was 24,883.2 miles long, while a modern estimate for the double meridian length is 24,859.868 miles. The latter is within one part in 3,200 of the ratio between the standard geographical Greek foot of 1.01376 (3168/3125) and 1.0125, the 81/80 foot (and that would make the Parthenon a "hundred footer" in its width). It is therefore reasonable to assume that between the building of

---

*Using the terminology developed by John Neal by 2000 and introduced in table 3.1.
†A day in angle on the equator was defined as being 360,000 feet, while the north-south mean degree (at 51 degrees) was 360,000 longer feet, namely Greek geographical feet of 1.01376 feet.

both Stonehenge and the Great Pyramid (by 2500 BCE) and the building of the Parthenon (designed by 447 BCE), an actual length for the meridian had been measured and that this had superseded the assumption (within the older metrology) that the meridian was half the length of the mean Earth circumference.

SANCHI GREAT STUPA, 232 BCE

Sanchi is a place in central India important to Buddhism; the Buddha visited there, and, following the construction of its Great Stupa, an important group of Buddhist stupas were subsequently built over 1400 years at the site. The Buddha was born in what is now Nepal in 562 BCE at a time when many new voices arose concerning the human potential for transformation. When the Buddha was breathing his last, he told his disciples to distribute his ashes, his "relics," and these could be placed within hemispherical stupas. The Great Stupa was one of reputedly sixty-four thousand of these containers made for Buddha's remains. In *Sanchi: Monumental Legacy*, M. K. Dhavalikar wrote, "The Great Stupa of Sanchi is the most impressive Buddhist edifice in India, and considering its antiquity and sculptural wealth, it is one of the grandest of man's creations [comparable] with the best monuments in the world."[6]

The best plan available to me was that of Frederick Charles Maisey, made in 1850. It shows a dome 105 feet in diameter and an annular perambulation area representing the sublunary world, as per the Equal Perimeter Model. Unusually, the dome's circumference, which has an elevated walkway around it, is 330 feet, or 3960 inches—the mean Earth radius, in miles, within the factorial version of the Equal Perimeter Model. Correspondingly, the outer circle of equal perimeter is 420 feet in radius, or 5040 inches, the distance between the two centers of the Earth and the Moon in that geometry, a distance corresponding to the capped height of the Great Pyramid in Sumerian feet (7 units to 11 on the base, hence $\pi = 22/7$). This important result suggests that the factorial version of this model is not just a later truth but was understood by the builders of the Great Stupa; only by using radii in inches could a stupa as small as 105 feet across represent the Earth in the miles as inches (a scale of 1/63,360).*

---

*This scale is to be seen again in John Michell's landform pattern called the Decagon of Perpetual Choirs. See the post "John Michell's Perpetual Choirs" on the sacred.numbersciences.org website in its Sacred Buildings section.

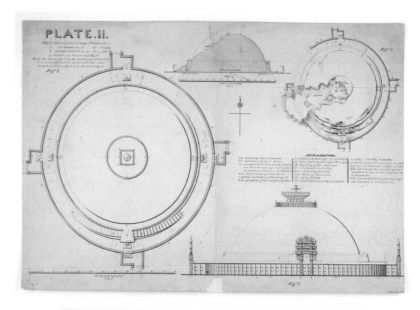

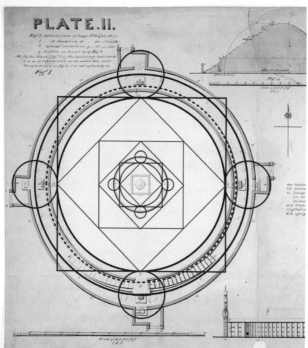

FIGURE 4.6. *Top,* plan, elevation, and section of the Sanchi Stupa by Frederick Charles Maisey (1825–92). *Bottom,* the Great Stupa interpreted as both Equal Area and Equal Perimeter Models, the latter making the stupa's dome a scale model of the Earth.

British Library, London, UK, copyright British Library Board.
All Rights Reserved/Bridgeman Images.

This means that a furlong of 660 feet must equal 7920 inches, which is numerically equal to the diameter of the mean Earth in miles. Every humble field, a plow's length long, reflects the idealized diameter of the Earth. How many Buddhists and others, when visiting this stupa, would know that the Great Stupa was a coded model of the primordial temple, the Earth itself a perfect sphere. And this seems to be the point, already mooted, that these monuments built by cultures and religions of all sorts did not need visitors to know the details of how they were subconsciously connected to the spiritual world.

THE HAGIA SOPHIA, 537 CE

Constantinople* was the largest and wealthiest city in Europe. Its vast Imperial Library contained the remnants of the Library of Alexandria and had over one hundred thousand volumes of ancient texts.† The defining unit used in the Hagia Sophia, in Constantipole (now Istanbul, Turkey), has been taken by Rowland L. Mainstone‡ to be 1.024 feet (ratio: 128/125) from the measurement of its dome: an intended 100 feet of 1.024 feet in diameter. This unit I will call the canonical Byzantine foot,§ equal to 128/125 feet and arrived at through 64/63 (the root foot) × 126/125 (see appendix 1).

The Hagia Sophia therefore has a dome with a diameter of 102.4 feet, which, considered as foundational, can be divided by 11 so that a unit length of 9.3̲0̲9̲ feet (512/55) is found, which is numerically half of the nodal period of 18.618 feet per year, in English root feet. We have encountered 18.618

---

*Emperor Constantine the Great moved the Roman capital to Constantinople in 324 CE. From the mid-fifth century to the early thirteenth century, the Byzantine empire was a major regional power until the Fourth Crusade sacked the city in 1204, enabling the Turks to establish, in 1453, the Ottoman empire in the city we know as Istanbul.

†Following "a series of unintentional fires over the years and wartime damage, including the raids of the Fourth Crusade in 1204 . . . the library was allegedly destroyed by the Franks and Venetians of the Fourth Crusade during the sacking of the city. . . . The majority of Greek classics known today are known through Byzantine copies originating from the Imperial Library of Constantinople." Wikipedia, "Imperial Library of Constantinople.

‡"The closeness of the length of the sides of the central square beneath the dome to 100 of these feet suggests that the sides were intended to be precisely this. If so, we obtain a more precise unit of between 0.312 m and 0.3125 m." Mainstone, *Hagia Sophia,* 177. 1.024 feet is 0.3121152 meters.

§In ancient metrology this module was unknown, and is here called the Byzantine module after the monuments and culture that used its standard canonical variant.

before as the magic *radius* of the Equal Area Model vis-à-vis a square area of side length 33 units. But here, something new seems to be afoot since a very good approximation to 18.618 is the ratio 1024/55, which equals 18.6<u>18</u>. If multiplied by 14, the diameter of equal perimeter to the square becomes 130.3<u>27</u> feet, for which the perimeter is 409.6 feet, which equals 22 lengths of 18.6<u>18</u> feet. The diameter is then fourteen of those units.

This would indicate the unit of 9.3<u>09</u> feet, one-eleventh of the dome's radius, was used to form a larger radius from the center of the church, 18.6<u>18</u> of these units, thus arriving at 18.6<u>18</u>² divided by 2, which in feet is 173.318, the exact numerical length, in days, of the time between eclipse seasons. This new radiant from the center is, when doubled to form a diameter, the eclipse year of 346.62 days in feet. This strange resonance of 18.618 as a number has been commented on without explanation, but we can see that it emerges from the nodal period of 18.618 years that defines the equal-area relationship to the solar years in 33 years. The nodal cycle is numerically resonant with the day length on the Earth, by which we mean the rate at which the nodes move 1 DAY on the Ecliptic, which is 18.618 days (see "The Third Triangle and the Golden Mean" in chapter 2). The

FIGURE 4.7. Lithograph of the Hagia Sophia.

area swept out by the nodes in 18.618 years equals the area of a square of 33 years.

The geometry emerging can be radically simplified by remembering that the eclipse year can be connected to the solar year through the three-square rectangle. A radius of 173.31 feet would need to be divided by 3 to get a side-length unit (for the triple-square rectangle) of 57.77 feet. The central region of the main open area of the church can form a cross 6-by-6 square units of this size (see fig. 4.9, p. 88, homing in on the smaller squares that can be seen in the cross). Eight radiating lines from the central point present the solar year as diagonals of eight three-square rectangles, visually resembling a Maltese cross.

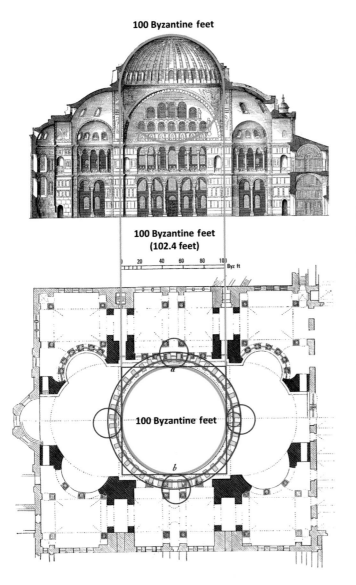

FIGURE 4.8. The Hagia Sophia dome from the side and, in plan, from above. There are forty supports, each 10 Byzantine feet apart, forming the sublunary annulus of the combined models of equal perimeter and, implicitly, of equal area.

The four large alcoves measure exactly 18.618 feet across, and their centers are half a lunar year in feet from the center, presenting the found unit 1024/55 = 18.618, which is inherently intimate with the Byzantine foot of 1.024 feet, whose root is 64/63 and whose standard variant is 56/55 feet.

The geometry of the octagon, albeit irregular, connects the nodal period and the cross motif with an interesting range of important synodic periods, as the octagon has vertices on the solar year diagonals and eclipse years between the cardinal directions. Since the lunar year as a diameter defines the centers of the alcoves, it seems appropriate to form a circle representing the Jupiter synodic period (purple) and the diamond of Saturn (black). Jupiter just touches the extreme corners of the 18.618-foot-diameter alcoves, and the form

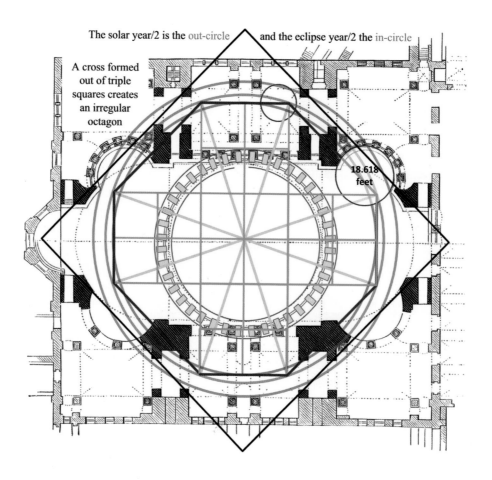

FIGURE 4.9. The formation of three-square rectangles that link the eclipse and solar years to build a Maltese cross.

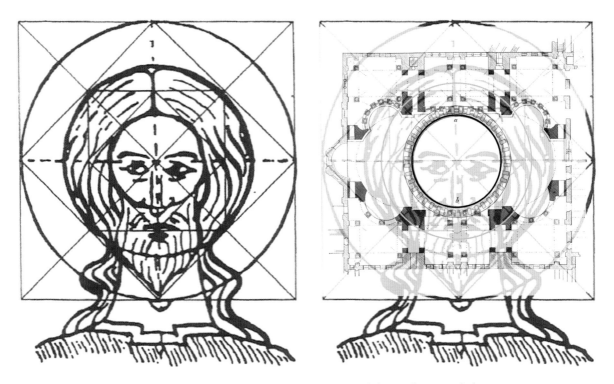

FIGURE 4.10. The Byzantine canon modeling Christ and the relationship of the outer diamond to the Hagia Sophia.

then resembles the Canterbury pavement (chapter 7), which could have originated in Byzantium before being reset by the pope.

The Dome of the Rock, built three centuries later in Jerusalem by Byzantine architects, used the figure of a *regular* octagon to express the Equal Area Model in the width of the walls, as was also done with the walls of the Kaaba (see chapter 8). The Dome of the Rock also used another ratio belonging to the inverse Byzantine module, whose root is 63/64, and a standard canonical 99/100 feet which, times 5/3, equals 1.65 feet, the Kaaba's ell. The nodal period probably expressed esoteric aspects of the Byzantine view of Christ. The cross can be made to show it, as can the Byzantine canon, used within the designs of icons (fig. 4.10)

REALIZING THE LAND

Larger geometries appear to have been created upon the land, and these are known by studying individual monuments relative to each other. In the

Americas, the Herculean efforts of tens of thousands of people were galvanized by some form of leadership to create Hopewell earthworks, Maya pyramids, and sacred cities like Teotihuacan, often aligned to the Moon's nodal cycle or the Sun's extremes of alignment. Where measurements and units of measure are found, these corresponded to those of the Old World, suggesting once again that metrology was a unified phenomenon with a global reach. The myth of Atlantis might support this as coming from America to the west (and to China and India), but also, the myth of Quetzalcoatl, the founder of Mexican sacred building and calendrics, says that he came from the Old World and had a beard and other foreign manifestations. The city of Teotihuacan undoubtedly used the MY (approximately 83 centimeters) as a unit of measure as well as the normally Egyptian Royal yard which, in its standard geographical form is also found at Stonehenge.

Akhenaten traumatically moved the spiritual capital from Thebes's Karnak temple complex to the geographical center of Egypt as defined along the Nile, four of whose degrees of latitude are represented by the Great Pyramid's four sides, that differ in length. To find the apex of metrology's development, one must turn to the early third millennium in dynastic Egypt. What had been started on the northern seaboard of Europe, as seen at Carnac in Brittany, in northwestern France, was completed when the resources of the pharaohs were applied to creating a spiritual model of the Earth itself. The megalithic techniques of triangles, alignments, and measures made the north-south valley of the Nile ideal for geodetic work. The same considerations concerning integers, commensurability, and ratios underlie the later mathematical texts of the Middle Kingdom, in which higher worlds could be represented on the Earth through sacred arts.

It would not be unreasonable to compare this descent of higher meanings onto the landscape as an *Arrival* (in the spirit of that 2016 film), in which preparatory work was necessary to establish cosmic standards as a foundation for human culture. Although they're now forgotten if not rejected, we still rely on many of the old symbols that became part of the human mind and soul-stuff. It was an intervention that was successful in creating conditions for human development. After the fact, this intervention involving metrology, buildings, and the planet's size, was overlaid by myths and legends while humans struggled to learn why they had been enlightened in the first place.

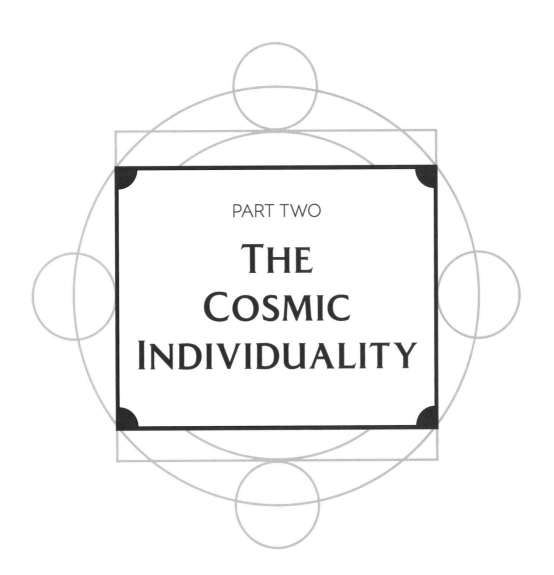

PART TWO

THE
COSMIC
INDIVIDUALITY

The universe is a manifestation of the Universal Will's desire for it, and this was described by the mystic and philosopher G. I. Gurdjieff as necessarily defeating entropy through perpetually renewing the higher worlds. Living systems have partially manifested this possibility through locally reducing entropy within their bodies and the immediate environment. Living systems are also *whole,* able to reproduce themselves and maintain a metabolism, while becoming more independent within the environment. At the current pinnacle of life stands the human form, able to move like animals yet also able to understand the world, change it, and reflect on it. With this comes our idea of selfhood, of an individual nature of one human vis-à-vis another. Yet this individuality is still largely potential because there are many group-soul characteristics inherited from animals, plus a penchant for delusion, egoism, and those deadly sins. But selfhood's potential appears relatively unlimited in scope.

Around 600 BCE, the human potential for self-development emerged via a number of different teachers who related the human self to the whole cosmos. In Buddhism, for example, the Buddha obtained a state of consciousness that was liberated from the limitations of materialism while remaining compassionate to all living things. That a human being can achieve such a degree of freedom* reflects the freedom of God relative to the universe. This is a theme within the Pythagorean notion of the microcosm and the macrocosm, where the human realizes that his or her own being is directly related to the structure of the universe, an idea found in the references to the cosmic within monuments and the smaller repeated patterns often found within them that imply the human imagination.

I have adopted British scientist and author J. G. Bennett's term Cosmic Individuality,† which describes a transformed state in which the individual finds that his or her own will and Universal Will are compatible yet renewed.

---

*Degrees of freedom are experienced when someone learns something about the world that radically enhances what he or she can do. Freedom then is not escape from doing but freedom and effectiveness in action.

†Bennett wrote a four-volume work called *The Dramatic Universe* that blended existing knowledge of the universe with ideas from Gurdjieff and others to recognize reductionist science as a great asset in forming what I find to be a new and much-needed cosmology.

This gives a name to an important goal for the universe: that human freedom might consummate the greater Purpose of the Universal Will, namely, enabling the universe to evolve new beings who can interact with the cosmic energies of consciousness, creativity, and love—the latter Bennett's unitary energy. In this view, angels are the manifestation of the Universal Will and responsible for the design and maintenance of planetary systems, while humans are initially lesser beings arising from a planetary world but with the potential to enter a world beyond the planetary system. This was called by Gurdjieff (see chapter 9), World 12, the cosmic level of our Sun, where Cosmic Individuality represents a transformed form of life and the fulfillment of life on earth.

This human potential was most clearly presented in the Abrahamic religions. The subliminal doctrine found in the early Bible was subtly transformed in the New Testament presentation of Jesus as the key synthesis between the Universal Will and Cosmic Individuality. Any outer differences between Hebrew, Christian, and Islamic versions of the Abrahamic religion are unified by the inner role musical tuning theory played, as a core concept that originated in 600 BCE Babylon, where the exiled Jews wrote down the earliest Bible story about the first man, Adam, whose name has a letter-number equivalence pointing to key limiting numbers for the musical octave, expressing the chromaticism (twelve notes) and hence the twelve patriarchal tribes like twelve chromatic notes.

MONOTHEISM AND THE MUSICAL OCTAVE

Traditional religions based their ideas on numbers in a way natural to the curiosity of the late Stone Age people who discovered special numbers that were later called sacred. That is, religious meaning was first born out of numbers, and religious symbols found their form from the apparent quantification of existence that numbers bring. Living forms were not predetermined to be the forms we call dog, beetle, and plant, but had to successfully exploit the natural environment in ways that are archetypal: "you are what you eat," and therefore "your body is how you find and eat it." Human ideas about the higher meaning of the world did not emerge accidentally but were informed by the early quantification of the *literally* higher time-world of planetary astronomy. The harmony of the spheres embodies ideal numerical ratios also found in music by the human ear and by the intellect, eventually giving precedence to the tones and semitones found between the notes, as scales within an octave.

Every child plays the intervals ready-made on an instrument without knowing any musical theory. The people of the ancient Near East of 3000–1000 BCE, despite their elite's prodigious grasp of musical ratios, instruments, and tuning methods, left scant record of the concrete notion that unifies our own music theory: the octave. For us, there is an octave of only twelve white-and-black notes, these asserting a primal authority in defining the meaning of a melody or chord and its key. The tonic* and its octave have become the root (or home note) of the octave's creativity. No concrete evidence has been found of ancient Mesopotamian music having had such an overruling perceptual rule.

The explicit primacy for the octave that appeared in first-millennium-BCE texts may have been monotheistic in origin. Around 600 BCE, the early Bible was being written, and Pythagoras was bringing Egyptian and other number mysteries to pre-Classical Greece. The origins of the numerical theories underlying these cosmologies and of the Mesopotamian and Egyptian mysteries before the first millennium BCE was astronomical, emerging from the astronomical civilization we call the megalithic. It was only in the fourth millennium BCE that musical cosmological ideas could first have been articulated, based on the tonal intervals existing between the recurring periodicities of the lunar year and the outer planets.

These harmonious interactions between celestial periods seen from the Earth, could be viewed as an ordering function for life on Earth and hence as being archetypal gods.† The legacy of this astronomical discovery, in the late megalithic, of musical ratios was therefore carried forward into the early civilizations of the third millennium BCE as the many gods. Only then did the significance of the octave, representing the oneness of creation and the significance of human selfhood, emerge as forms of monotheism in the first millennium BCE.

---

*An octave is named after the eight notes of the octave, one being the tonic, which then repeats as low and high versions of itself. The tonic, or *do,* which is doubled in frequency when ascending or halved when descending, is heard as the same note. Any one of the twelve note classes can become the tonic or key of the octave experience.

†We now know that planets subsist on a numerical substrate of friction-free gravitational dynamics—they are perpetual motion machines.

A MUSICAL ARRIVAL

The melodic music of the ancient Near East gave voice to the music of the planetary gods, wandering like medieval plainsong over a world of tones and semitones colored by the subliminal influences of the octave's tonic.* The ancient and popular form of five notes (*do-re-fa-sol-si-do*) was a music without much trouble; today, five-note Pythagorean wind chimes are played by the wind to marvelous effect. In this and other ways, the popular music of the ancient Near East appears to have not queried the numerical nature of music as a holistic structural form, but more simply saw its inevitable arising within physical strings according to their dimensions when measured by their metrology.

The Mesopotamians became quite obsessed with divination and astrological portents, which are configurations caused by the interaction of celestial periods. One cannot assume that our own religious or musical theories were present. It was the Sumerians who first had gods who created the world musically, but these died, to be replaced by others. The Flood-hero Marduk cut up Tiamat and her retinue of ill-disciplined and inharmonious gods, leaving the harmonious intervals that define harmony.

The very making of theories is now thought to be a Greek development, before which the civilized world was organized by scribal recipes. Pythagoras, returning to Samos from the ancient mystery centers in his fifties, forged a new cosmological theory based on numerical tuning. In his myth, inspiration for a spiritual cause in numbers came from Babylon and Egypt, and his subsequent musical experiments famously varied string lengths on a monochord.

Pythagoras developed a pyramidal array of numbers called the Tetractys, in which the first ten numbers work their way downward from the number 1 at the top (see fig. P2.1, *left*). An adaptation of the Tetractys (fig. P2.1, *right*) is the Lambda diagram, where the left-hand radiant from the top holds the powers of two, of which one is the zero power.† This radiant was complemented by a right-hand radiant of the same powers, but this time powers of three.

---

*The tonic affects music as a center of gravity that can move, affecting the other notes by making them sound different in a new context.

†Since 2 to the power zero is 1, as is the zero power of all numbers.

The Tetractys

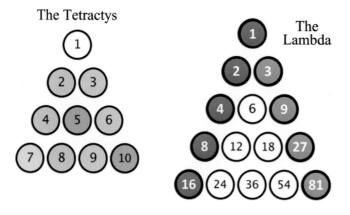

FIGURE P2.1. Two
powerful diagrams of
the Pythagorean
number arrays:
the Tetractys and
the Lambda.

The numbers 1, 2, and 3 were atop both diagrams, prefiguring their role in the musical creation. The Lambda was very important for his numerical tuning theory since every cross-multiple of the powers of two and three can be found in one location only, as the Lambda continues downward. The numbers between these two radiating lines of powers are their interaction as multiples of both 2 and 3 and their higher powers (called exponents). The bottom row of this Lambda diagram is {16 24 36 54 81} and each number to the right is 3/2—a musical fifth—relative to its left-hand neighbor, and Pythagorean scales are made up of notes separated in this way, by fifths. In figure 9.5 (p. 213), the five-note pentatonic scale has a central tonic of {36} (fig. P2.1), quadrupled to 144 so that it can be greater than 81. The octave becomes {128 96 (72:144) 108 81}, the repeated tonic being the octave doubling from 72 to 144. So, the Tetractys can grow ever wider rows, can create an extra Pythagorean note and, by the seventh row (not shown), the seven notes define the Pythagorean heptatonic, making this Lambda diagram a cunning demonstration of musical tuning, using numbers.

Pythagoras is said to have discovered that when one vibratory tone is increased by a power of 3, a new and different note is created. After four powers of three, a pentatonic musical scale is created. In this pentatonic, called so by having five different notes, and -tonic meaning "within an octave," an octave is created when the first note (the tonic) is doubled to also end the scale. The new tonic appears in the center of the new row of numbers, which has balanced powers of 2 and 3, as with 36 which is both 6^2 and 4 times 9.

The Lambda diagram automated the creation of suitable numbers for the creation of Pythagorean scales, and its "creative power" was shown in the

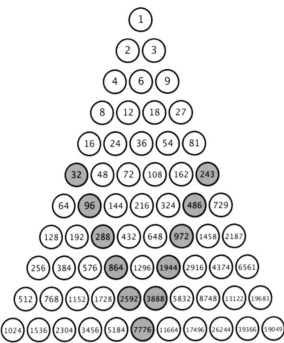

Apollo keeps the harmonic development of
earth within his limit of 7,776

FIGURE P2.2. The natural limit of balanced powers of two and three, five powers of each percolating down (in a *V* shown shaded) to the combined limit of 7776 associated with Apollo and YHWH as 777,600,000—the harmonic limit of the geocentric planetary world.

For more on this association, see Heath (2018), 67–69 and 147–48.

numbers 1, 2, and 3 atop the two diagrams—the starting tone (1) affirms a tonic for an octave (2) of high and low *do,* which is then penetrated (3) by successive powers of 3 to create new notes within an octave.* Creating a first new note via an ascending interval of 3/2 shows the new note to be related to the tonic as 2 while being three halves of the tonic. A created note belongs to the note from which it was tuned, the ratio 3/2 being inherently tied to 2, the octave. Eleven successive tuning operations populate an octave of 12 tones relative to the first note, and in the Greek context the central number is 7776—the total number of syllables in Plato's *Ion* in which it is asked "And where is Apollo in *Ion?*" the answer being hidden in the number of its syllables.

If more successive tunings are performed, out-of-tune versions of the same note class (Tiamat's disharmonious brood) occur due to the Pythagorean comma: for twelve fifths (of 3/2) can never exactly equal seven octave doublings. The ear demands that the octave be equal to 2, but the powers of the prime number 3 (on

*The number 3 can divide into 2 to produce fractions that are different notes.

the right of the Lambda diagram) cannot be removed to leave only a power of 2.*

The solution to this problem is visible in the two diagrams in figure P2.1: the Lambda has the number 6 beneath the number 1, while the Tetractys has the number 5, the next *prime* number after 3. The creators of ancient Flood heroes understood that by moderating powers of three with powers of five, the detuning inherent in Pythagorean octaves with more than seven notes, tuned using 3/2, could be moderated by the two new semitones of 16/15 and the two new tones of 10/9. These new ratios remove the powers of three while cancelling their own powers of five in the process. This is what originally created our modal scales, named after the Greek tribes. The key factor is 81/80, the synodic comma already installed as the ratio between the lunar year and the 30-sidereal-day month. The new semitones and tones differ from those of Pythagorean tuning by this synodic comma; for example, the Pythagorean semitone of $256/243 \times 81/80 = 16/15$ becomes enlarged and more harmonious.

In the sky, Zeus-Jupiter delineated his pantheon of just twelve gods and a Zodiac of just twelve signs,[†] borrowed perhaps from the agrarian calendar of Mesopotamia.[1] The Indo-European myths of Greece and India carried this sacred association, which is found also in the twelve disciples, regions, gods, hours of daylight, and musical note classes, with the move toward a monotheism having discovered the unity in twelve notes, a relic of the numbers implicit within musical ratios echoed in the twelve months in a lunar year.

By the twentieth century, tuning based on numbers was replaced by equal temperament of twelve semitones of equal but irrational size. Monotheism had mutated through the new dispensations of Christ and Muhammad, each a unique religious vision having new hidden meanings that were often based on numerical, astronomical, and musical invariants then considered heretical by their ecclesiastical authorities. The inner story of the power of the octave to provide a unified God and creation provided the outer story of the Bible. And its inner numerical metaphor of numbers preserved an inner doctrine of astronomical harmony surrounding Earth.

---

*Prime numbers are indivisible by any other number (1 not being a number but the oneness each number expresses). The interaction of early prime numbers has been key to my understanding ancient numeracy and its available methods.

†Jupiter travels through the 30 degrees of a single Zodiac sign in $19 \times 19 = 361$ days, which is an excellent reason for that planet to be associated with the number 12, and, indeed Zeus, the Greek Jupiter, had a pantheon of twelve gods.

5

Harmony of the Local Cosmos

LISTENING TO STRING RESONATORS, it is possible to discover that music arises from ratios that involve small number differences between their lengths, usually only one unit in difference, making ratios that are called superparticular—such as 3/2. All things being equal, this reveals numbers as creative through those ratios and leads to a tuning theory like that of Pythagoras, in whose myth musical theory only then becomes extrapolated into a theory that the cosmos is harmonious. But this story is back to front since, long before Pythagoras, the megalithic astronomers had compared cosmic time periods instead of musical strings. The cycles of the Moon and the outer planets displayed ratios between them that were then seen as uniquely connected to the acoustic music made using strings: lengths of time already had units of length, and this direct way of storing and manipulating numbers developed into metrology. Both astronomy and acoustic experiments revealed musical ratios as belonging to a special domain of harmony that is factual while also expressing the value system we call harmony. Harmony evidently also exists within the human soul, since music can be heard as well as be understood intellectually, through numerical ratios. The hearing of musical ratios is fundamental to how one *can* recognize a harmonious interval at all. Through comparing time cycles, no sound is directly heard, but string lengths of similar ratio can give them voice. And while an instrument is required to hear music, a cosmic instrument had been made of the planetary world.

Numerical tuning theory of some sort has existed since 3000 BCE. While the full scope of early tuning theory can be argued over by musicologists, it probably emerged alongside a metrology to build and count lengths. Once known, the musical ratios between the lunar year and the synods of the outer planets would encourage the study of stringed instruments such as harps and lyres, which are based on the playing of many different string lengths. Metrology provided smaller units of length than the foot, such as digits, inches, and fractions of these, which functioned like millimeters do today, in defining string lengths.

Through varying string lengths, number relations could be established between harmonious notes as interval ratios, in which the human ear and intellect then established an integrated tuning theory: if the string was 60 in length, then blocking it off at 40 and 45 would give the fifth and fourth tones of the scale, while 30 would give the octave relative to the full string length. A block at 50 would express the major third (between 40 and 50) and the minor third (50 to 60). The Sumerians called their primordial gods by the numbers 60 (Anu), 50 (Enlil), and 40 (Ea-Enki), and the tuning texts for harps, found on later cuneiform tables, appear to call the strings of harps by the smallest-possible tuning lengths for the given notes relative to each other, implying their tuning theory was numerical. Their harps had to be of a certain size to form the musical scales that the instruments were able to accommodate, thought to be enneatonic with two overlapping pentadic scales and no obvious evidence of the octave as such. Since the scribes only wrote what they were told, someone in the chain of players, tuners, instrument makers, and possibly priests knew numerical tuning theory.[1]

The simplest model of harmony is to be found in the first six numbers, since the ratios between these adjacent numbers give us the five larger intervals of music as pure tone ratios of octave, fifth, fourth, the major third, and the minor third (1:2:3:4:5:6) or {unison octave fifth fourth major-third minor-third}. These numbers sum to 720, the number forming the unit in the model of equal perimeter (chapter 3) or, as we would say, factorial six (6!). Pythagorean musicologists of recent times called the first six numbers *senarius* (meaning "born out of six"), a term more familiar as a meter for Homer's epic verses (fig. 5.1).[2] The senarius can be found expressed in the rectangular floor plan of the Kaaba (see table 8.1, p. 193), but without seeing its deep history, specific instances might be thought to be simple flukes.

The number 720 was associated with the Patriarchs of the Bible (which was written down c. 600 BCE). The Patriarchs descended from Adam (A. D. M), whose name, in the number-letter equivalence called gematria, represents

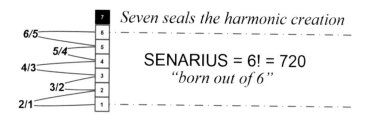

Seven seals the harmonic creation

SENARIUS = 6! = 720
"born out of 6"

FIGURE 5.1. The senarius of the first six numbers
terminated by the seventh (*in black*).

1 + 4 + 40 = 45, which is 720/16. From this 45 the genealogy of the Patriarchs
emerged through doubling Adam, such that Sarah, wife of Abraham, gives birth
at age 90 and then Isaac lives for 180 years.

The number 720, which is 4 times 180 (Isaac), is harmonically significant because at that limiting number for an octave, the numbers in the octave
360:720 allow five Just musical scales* to appear *automatically,* making 720 a
key number within the model of harmony. And, when 720 is multiplied by 2,
1440 is the number used within the Sun-Moon-Earth system of time to ensure
a harmonious relationship between the Moon[†] and the inner and outer planets
as well as many other phenomena, such as the sidereal day, 30 of which are
81/80 of the lunar month, the interval called the synodic comma.

As already stated, the sum of the senarius is also seen in the factorial
extension of the Equal Perimeter Model (see fig. 3.1, p. 60). One sees (with
Michell) that 6! = 720 can be multiplied by the units of that model to make
it correspond to the actual sizes of the mean Earth and the Moon in miles.
That is, this model seems to have been employed as a template for their relative diameters to be 11 to 3. The volume of the original Earth and its collider
must have been divided accordingly, billions of years ago, while also defining

---

*In modern terminology, these five scales are called Dorian, Aeolian and Mixolydian, and Phrygian
and Ionian, the latter two pairs being like twins as they are the inverse of each other—each one
rising through the other's descending order. This symmetry occurs due to the diatonic structure, in
which the two semitones always lie opposite one another in the octave's tone circle (see my book
The Harmonic Origins of the World, fig. 5.10).

†The moon has 12 lunar months in its year and if each of these is seen to be made of 80 equal
parts, then the lunar year is 960 units. The octave is then 720:1440 with the lunar year as fourth,
Jupiter (1080) as fifth, and Saturn (1024) as tritone. The octave is (9:12::13.5:18) lunar month, a
fifth (3/2) greater than Plato's World Soul of (6:8::9:12).

the angular momentum of the Earth's rotation and a lunar orbit evolving from close to being the size of the Sun from Earth.

When the Earth diameter of 11 units is multiplied by the senarius of 720 miles, it becomes 7920 miles, which is also equal to the quarter circumference of the circle of diameter 14. The Moon is then diameter 3 × 720 = 2160 miles. The mile is a subunit, being the diameter of the Earth divided by 11 × 720 (the 11 of the model and the 720 of the senarius) so as to be a harmonic unit with respect to the Earth. By this means, the Earth and the Moon (a) have the senarius built into their absolute sizes in miles while (b) conforming to the pattern of equal perimeters. This was already clear earlier: the Great Pyramid, with its height of 7 and base of 11, also has the factorial model beneath it, giving a good first approximation for the size for the Earth (radius 3960 miles) and the Moon (radius 1080 miles).

THE HARMONIC PLANETARY MODEL

In *The Harmonic Origins of the World,* I exhaustively deal with the harmonic model, so here, the focus is on the harmonic model at work within monuments and elsewhere. The starting phenomenon, required to realize the harmonic model in the first place, was megalithic quantification of synodic periods, those of the outer planets expressing musical ratios with respect to the already quantified lunar year.

Integer ratios in the world of time seem quite irrelevant to modern celestial dynamics since science uses sophisticated Hamiltonian functions, which involve advanced calculus to study orbital dynamics. Yet the integer relationships between celestial bodies allowed a simple numeracy, using geometrical and metrological methods, to establish a very intelligible model of the planetary world of time. The extreme simplicity as ratios of 9/8 (Jupiter) and 16/15 (Saturn) was a discovery of a higher level of order than the preliminary megalithic discoveries of the nonmusical ratios between the time cycles of the Sun, Moon, and Earth. Human music making was likely to have flowed from this discovery long before the Pythagorean acoustic experiments and perhaps before the Sumerians detailed their own numerical tables for the tuning of stringed instruments—since they left no written account of its astronomical significance.

When the Sumerians brought new arithmetic techniques, they created harmonic tables involving very large numbers in base-60, 60^4 being used as a limit. Numbers less than the *fourth* power of 60 (12,960,000) were known to Plato, who called this number a "sovereign number."[3] This very large but harmonic number (60^4), belongs to the inner planet Venus within the matrix of numbers based upon dividing the lunar month by 80 units. Venus, known as Inanna/Ishtar in Mesopotamia, was the primary goddess of the ancient Near East, and she was called Quetzalcoatl by the Olmec and Maya of Mexico.

By 600 BCE, Venus as goddess was being displaced by the monotheistic Bible writers, whose YHWH (6.5.10.5) has an exponent value of the *fifth* power of 60 (777,600,000). This is, I have found, to be the limiting number of the entire system of time on Earth (see fig. P2.2), and in Genesis, chapter 2, YHWH creates Adam (A + D + M = 1 + 4 + 40 = 45), who is also 1440 when viewed in position notation. As mentioned above, the matrix value of the octave between 720 as 9 lunar months and 1440 as 18 lunar months reminds us of Olmec and Maya supplementary glyphs, the 18 months that were sometimes appended to their Long Count calendar. The Canterbury pavement (see chapter 7) expresses the octave as a 9-foot square for the microcosm and around that an 18-foot extent for the macrocosm—this showing the synods of Jupiter (13.5 feet) and Saturn (12.8 feet) relative to the lunar year (12 feet).

Though megalithic astronomers could look at the sky, their measurement methods were only accurate using horizon events. Horizon observations of the solstice sunrise and sunset each year and of lunar extreme moonrises or settings (over 18.6 years) allowed them to establish the triangular and metrological ratios between these and other time periods, including the eclipse cycles. As stated in chapter 2, triangular models with trigonometrical features evolved into a metrology in which integer ratios of all sorts functioned like the two longest sides of a triangle.

Having counted the lunar and solar years to establish the Lunation Triangle (simply the geometry of four squares), it was natural to compare the periods between the synodic loops of the outer planets (fig. 5.2, p. 104). The synod of Jupiter can be measured by observing its loops in the sky, against a backdrop of stars, in which Jupiter heads backward each year, this as the Earth passes between Jupiter and the Sun. Jupiter is said to go *retrograde* relative to the general planetary direction toward the east. Since such retrograde movement occurs over 120 days, Jupiter will set 120 times on the horizon

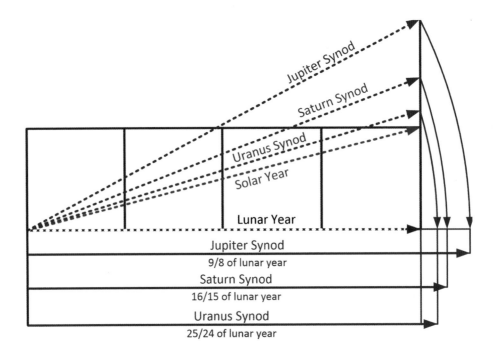

FIGURE 5.2. The synods of the outer planets in the context of
the geometry of solar and lunar years, that is, the Lunation Triangle,
and equally the diagonal of four squares.

while it is moving in retrograde. This allowed megalithic astronomy to study
the retrograde motion of Jupiter and observe *when the Moon was in conjunction with Jupiter* in the night sky, that is, when the Moon set with Jupiter's
own setting (fig. 5.3).

This method exploits the fact that Jupiter reaches its *maximum* retrograde
motion halfway in the loop, 60 days after its initial standstill in the sky. If,
at that point, there is a conjunction of the Moon and Jupiter, *then the Moon
must be full:* the Sun will already be opposite Jupiter, and so the Moon setting
with it must be full. Thus, when a full Moon is in conjunction with Jupiter at
mid-retrograde, it will set to the west, and one can start counting lunar months
until the same phenomenon occurs. We know that counting moons soon displaced the counting of days because of the Moon's convenient integer relationships with significant long-term periods, such as the Saros period (223 months)
and the Metonic period of 19 years (235 months). Megalithic monuments could
predict the punctuation of these longer cycles of the Sun and the Moon because

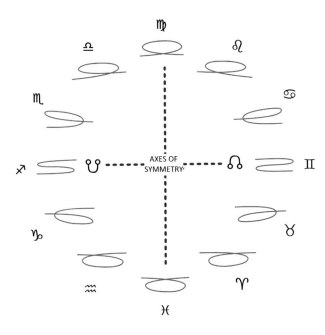

FIGURE 5.3. The
metamorphosis of the
loop shape when Jupiter
is opposite the Sun in
different signs of the Zodiac
(after Shultz, 1986, fig.131).

of the eclipsing phenomenon, which requires a near exact full or new Moon, respectively, opposite or in conjunction with the Sun.* The musical interval of the Jupiter synod (13.5) to the lunar year (12) was similarly simple to observe by counting lunar months because the Sun is opposite Jupiter at the middle of its retrograde loop. We know a single synod of Jupiter is 398.88 days, and so there are *exactly* 13.5 lunar months in each synod since 398.88/29.53059 = 13.5073, just over 5 hours longer than 13.5 months.

It is therefore true that if one counts between full Moons occurring 60 days into successive retrograde loops of Jupiter, it will be two whole Jupiter synods before a full Moon occurs in the same visual offset to Jupiter at its maximum retrograde in the sky. The lunar counting in-between will be 27 whole months, and at that point, the synod can be known as half of that, as exactly 13.5 months long and 9/8 times longer than the lunar year of 12 lunar months.

In this way, the length of the Jupiter synod was easily counted in months, the native unit of the harmonic matrix in the context of megalithic horizon astronomy. Knowing its length in months relative to the lunar year revealed the remarkable harmonic ratio of the Pythagorean whole tone between the Jupiter

---

*And we know that at Carnac and in the later megalithic period, astronomers moved to counting months using MYs since it was often simpler and gave important results.

synod (9 units of 1.5 months) and the lunar year (8 units of 1.5 months). This would have introduced the megalithic period to that *uniquely* simple category of ratios responsible for the highly ordered world of musical harmony. Note also that Jupiter's units of 1.5 months is 3/2 of the lunar month while the Jupiter synod is 3/2 of 9 lunar months.

The same procedure applied to the Saturn synod would require 5 Saturn synods (378 days) to complete since only then do 12.8 lunar months per synod yield an integer number of 64 lunar months. The loop of Saturn is smaller than that of Jupiter, 6.5 degrees compared with Jupiter's 10 degrees. The reverse is true, though, of the days spent by each planet in its loop, Saturn taking 140 days rather than 120. The ratio of the Saturn synod to the lunar month is another crucial musical interval, the semitone of 16/15. Saturn would also be found with a full Moon, 70 days into the loop, after 64 lunar months had passed from the last full Moon at that retrograde point.

Tones and semitones are so crucial to the formation of musical scales that megalithic astronomers, once aware of these intervals, would have started to investigate practical music and the stringed instruments that make musical intervals. And whereas the strings vibrate when plucked, the vibration counted within a celestial period was the daily rotation of the Earth or the lunar month, both relative to the Sun. The story of Pythagoras preserved this deeper historic origin for numerical musical theory, discovered through astronomy.

REPRESENTING PLANETARY HARMONY

The numerical simplicity of the planetary world derives from the fact that the inner and outer planets all have synodic periods relative to the same phenomenon—the lunar year. All the harmonic intervals relative to the lunar year are governed by the products of the prime numbers 3 and 5, which are able to penetrate the octave of 9 to 18 lunar months. The two outer planets are, in common, distanced from the lunar year by 9/8 and 16/15, but how can this be understood as a pattern within an octave defined by the prime number 2?

In antiquity, the planets were gods, and in the Sumerian version, the top god, Enlil, belonged "up there" and was allocated the number 50 (2 × 25), indicating that the powers of five run upward while the powers of three are horizontal, defining as they do the cycle of fifths, which acts like a spine for numerical tuning. Another mythic vision is of mountains and gods, and in this respect, one can shift the powers of five to run upward at 45 degrees so as to straddle two numbers below and form a very exact "holy mountain" of numbers based on the three numbers of the First Triangle {3, 4, 5} since 4 has no threes or fives.

Primal Grids to Define Octave Limits

The method used involved separating the prime numbers within such ratios. For example, as mentioned, Pythagoras developed a way of looking at the numbers found in musical ratios with his Lambda diagram (fig. 5.4).

The diagram on the left, can locate the 9/8 ratio of Jupiter's synod to the lunar year (rows 4 and 5 down, 24 to 27) but not the 16/15 ratio, since 15 involves the prime number 5, not included in this diagram. The obvious solution, on the right of figure 5.4, in the absence of a three-dimensional diagram, is to subtly collapse three dimensions into two using the fact that a hexagon looks like a cube seen through one of its opposite diagonals.

This approach, rediscovered by music professor Ernest G. McClain when

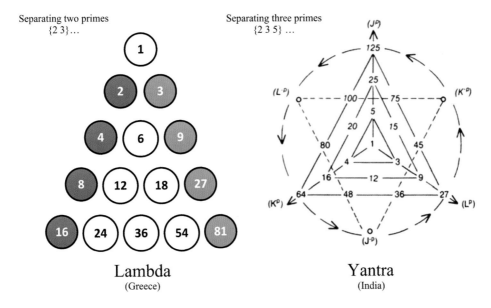

FIGURE 5.4. (*left*) Pythagoras's Lambda diagram of primes 2 and 3 as to their combination when cross-multiplied. (*right*) View of the first three prime dimensional products {2 3 5}, obtained by collapsing three dimensions into two which does the same to capture extra tuning ratios known as Just intonation.

Image on the right from McClain, Myth of Invariance, *1976, chart 10.*

reconsidering Plato in his 1978 book, *The Pythagorean Plato*, implies that Plato understood ancient tuning theory and its methods but had written in a cryptic style for initiates who already understood these matters. Fortunately McClain, a Pythagorean himself but also a wartime cryptologist, was able to read Plato's cipher.

The Holy Mountain of Celestial Harmony

Since we know of tones and semitones through our use of musical scales and that, in theory one can make a set of numbers in which the ratios required for a scale are all available, McClain realized that the ancient method was much more holistic and employed the method of setting numerical limits in which scales then automatically arose within a single octave's limits so that all the products of only the odd primes {3 5} could be graphed as powers, as possible tones within a given octave's numerical limits. Using this approach, one can discriminate the Saturn synod from Jupiter's, as it involves a negative (reciprocal) power of 5 seen within the denominator of 16/15 relative to the Moon, namely 15, which is 3 × 5. Saturn has no prime 3 factor at all, just powers of 2, the Moon's single power of 3 being removed by the semitone.

These synods can therefore be seen in relation to the lunar year (12 months) and the octave (9:18 months), in figure 5.5: 12.8/12 = 16/15, and 13.5/12 = 9/8. This sort of model was probably known but undocumented in the ancient Near Eastern period, when arithmetic was displacing metrological methods of

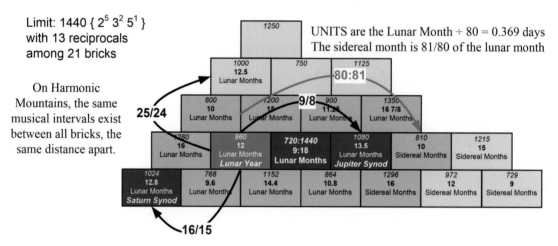

FIGURE 5.5. A matrix of Adam's octave of 720:1440 in matrix units and also numbers of lunar or sidereal months. At the top, the hardly visible planet Uranus is the brick for 1000 (12.5 months).

calculation—making such calculations possible. The vectors shown in figure 5.5 between the numbered "bricks" are the same between any bricks on the mountain when the same distance and angle apart. Therefore, such "holy mountains," as McClain called them,* were full of musical intervals, though some are quite obscure.

THE MEGALITHIC MODEL OF PLANETARY HARMONY

The directly triangular modeling of figure 5.2 appears to have advanced by the time of the British stone circles that started to displace earthworks (the henge, the cursus, and the long barrow) around 3000 BCE. An unexpectedly significant example was found in a *flattened* stone circle in Cumbria called Seascale (a.k.a. Grey Croft; fig. 5.6), the design of which was called Type D by Alexander Thom, who probably surveyed it in 1955. Only two of that type are known in the megalithic world, and I believe it would have allowed the Sun, the Moon, and the outer planets to be counted efficiently through using the Lunation Triangle within a four-square rectangle set out within it. I have also found that the Moon's nodal cycle of 6800 days could have been counted around its perimeter because of its unique dimensions in Thom's value, at that time, for the megalithic yard: 2.72 feet contains the prime number factor 17, as does 6800 days and 12,053, the 33 year cycle of the Sun.

Such flattened circles, when small enough, were probably made using ropes and stakes to first create a circle. One part of the circumference was then shortened to make a perimeter closer to three times its diameter. Thom thought the builders might, in fact, have been trying to reduce the irrational ratio π (which defines diameter to circumference dimensions) from 3.1416 . . . to an integer 3. Thom seems to have ignored historical metrology, probably because the megalithic was prehistoric, and besides, he was focused on finding what measures were most likely used—with an open mind. This meant Thom could not guess the builders had better ways to manipulate π using metrology: for example, using 176/175 to maintain whole numbers in both radius and circumference.

---

*Because of their use, in his view, in forming harmonically inspired narratives within works such as the Bible or Homer's epics.

Another take on the purpose of flattened circles was Robin Heath's observation that a Type B geometry generates a symmetrical pair of the Third Triangle (found in three-square rectangles) within it.[4] It was not too unlikely therefore that Seascale, a Type D flattened circle, should have a pair of four-square triangles as its *defining purpose,* probably constructed before the stone circle. In chapter 2, the Lunation Triangle was introduced as the vital geometrical model that gave megalithic astronomers mastery over the ratio between the duration of a solar year and a lunar year, allowing the solar and lunar calendars to be co-related numerically. While this triangle was originally arrived at through day-inch counting at the Le Manio Quadrilateral (see chapter 1), the observation that its third side was exactly one-quarter of its base enabled four-square rectangles to accurately reconstruct the Lunation Triangle between the rectangle's long side and its diagonal.

There are many confirming features within Thom's survey (fig. 5.7) that point to this as an ideal geometrical model. The major diameter (MN) in the survey could equally well have been 90 feet, rather than 89.9 feet. The flattened diameter (AB) is 81.6 feet, correct for the lunar year as 12 megalithic

FIGURE 5.6. The Seascale stone circle as seen from the south, with a nuclear reprocessing plant behind.

Photo by Barry Teague.

Seascale stone circle as 4-square Counter

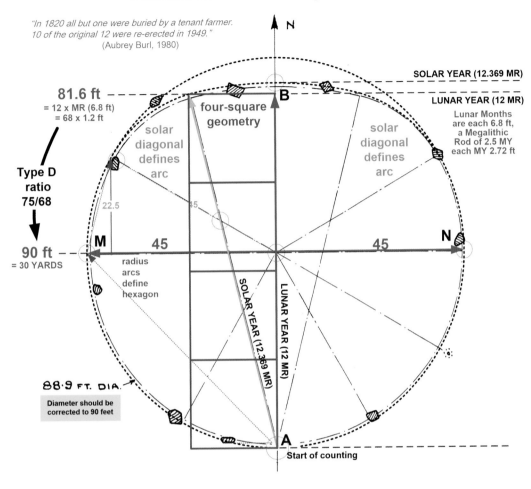

"In 1820 all but one were buried by a tenant farmer. 10 of the original 12 were re-erected in 1949."
(Aubrey Burl, 1980)

N

SOLAR YEAR (12.369 MR)

81.6 ft
= 12 x MR (6.8 ft)
= 68 x 1.2 ft

LUNAR YEAR (12 MR)

Lunar Months are each 6.8 ft, a Megalithic Rod of 2.5 MY each MY 2.72 ft

B

four-square geometry

solar diagonal defines arc

solar diagonal defines arc

Type D ratio 75/68

22.5

45

90 ft
= 30 YARDS

M

45

45

N

radius arcs define hexagon

SOLAR YEAR (12.369 MR)

LUNAR YEAR (12 MR)

88.9 FT. DIA.

Diameter should be corrected to 90 feet

A

Start of counting

FIGURE 5.7. Seascale's geometry could facilitate two four-square Lunation Triangles and outer planet synodic periods outside the limits of the circle.

*Underlying Seascale survey by Alexander Thom, c. 1955,
published as Thom, Thom, and Burl, "Megalithic Rings," 1980.*

rods of the later British variety (2.5 × 2.72 feet), each equal to 6.8 feet. The flattened diameter then fits the geometry as a four-square rectangle whose diameter *AB* (then the solar year) defines an arc that bridges between the forming circle of diameter *MN* equal to 90 feet and the arc of the solar year length, to the north.

The British archaeologist Aubrey Burl tells us that "in 1820, all but one [stone] were buried by a tenant farmer. 10 of the original 12 were re-erected in 1949."[5] Hence, can one believe that the stones are close to their original positions? A ground survey with geophysical equipment might be able to see the old sockets, but the use of megalithic rods within a four-square rectangle reveals a wealth of metrological and calendrical utility within this geometry at this exact scale, pointing to the dimensions of the stone circle being true to those when it was first built.

A recent analysis of Thom's survey for the Clava Cairns, near Inverness, Scotland, revealed a special length of 204 feet that would result from a count of mean solar months over the nodal period of 81.618 years (6800 days), if and when Iberian feet were employed to count days, and 27 feet to count months, as at Crucuno (see fig. 1.5, p. 26). These 223 mean solar months, each reduced to one Iberian foot for counting, equal 204 English feet in length. Fifteen megalithic rods equal half of that length, or 102 feet (15 × 6.8 feet). This means that the addition of *a further square* to the four squares, on the north, would define half the nodal period (3400 days) south to north.

Unexpectedly, a small type of digit called shu.si, known to historical metrology, can count the days within this 104-foot length to give half the nodal cycle. This shu.si (or "finger") of 0.36 inches historically divided the Assyrian foot of 9/10 (0.9) feet into 30 shu.si, and the double foot into 60 shu.si, compatible with the base-60 arithmetic of Mesopotamia. At just 3/100ths of an English foot the shu.si divides 102 feet into exactly 3400 parts, to numerically become the day-shu.si count over half of the nodal period.

The megalithic rod Thom found was 6.8 feet long, and the denominator of the shu.si is 100, which generates 680, which in 15 (3 × 5) megalithic rods gives the numerator of 3, while the remaining 5 (in the 15) then turns this 6.8 into 34, resulting in 3400 shu.si. This new unit seems to have emerged to serve the relationship of 15 megalithic rods to 102 feet when *an extra square is added to the four squares* (fig. 5.8).

It now seems likely that the MY of 2.72 feet, found by Thom within the megalithic geometries in Britain, had naturally emerged (with no error) from counting the Moon's nodal period: its fractional nature is 272/100, and 272 divides by 17 six times and by 34 three times, so that, when multiplied by 5/2, the rod is 68/10 feet long. This combination of factors means that Seascale was part of a greater scheme, as per figure 5.8.

In developing portable observatories, within which counting is made reliable, the four-square extension to five would allow the nodal period to be counted to and fro, between successive minimum and maximum standstills of the Moon. It is also true, though, from the point of view of the harmonic model, that a sixth square could also be added since that would represent the 18 lunar months (of the Maya's supplementary glyphs) that are at the top of the octave containing the Moon (12), Jupiter (13.5), and Saturn (12.8). At such

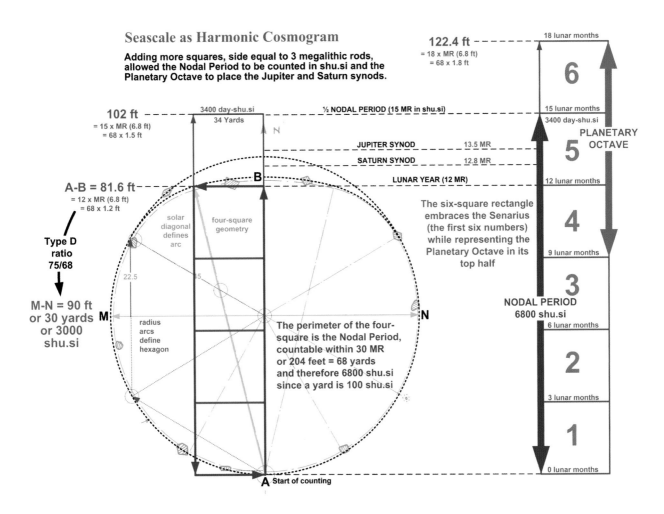

FIGURE 5.8. Seascale with additional squares added to embrace the nodal period and planetary octave.

Underlying Seascale survey by Alexander Thom, 1933, published as Thom, Thom, and Burl, "Megalithic Rings," 1980.

a point, one also has the senarius of the first six numbers {1 2 3 4 5 6} that is Plato's World Soul (6:8::9:12), only enlarged by a musical fifth (9:12::13.5: 18). In this model, it would be easy to place the synods of the outer planets "beyond" the lunar year, remembering that each megalithic rod contained 100 (2.5 × 40) megalithic inches.

Further inspection also reveals that the nodal period could be tracked in two different ways.

- First, the 3400 shu.si of the five-square extension could be counted to and fro.
- Second, the perimeter of the four-square extension will contain 30 megalithic rods, twice 15, equaling 6800 day-shu.si. Therefore, the perimeter could count the whole nodal period without the extra fifth square. In all probability, therefore, the fifth square was unnecessary and the notion of the octave cannot be assumed to have existed, but the 12.8 megalithic rods and the 13.5 megalithic rods of the months in the Saturn and Jupiter synods could have been marked by stones outside the flattened portion.

One can see from the above how a flattened circle of Type D could have functioned as a general purpose astronomical framework within a conveniently built form (fig. 5.8) rather than being laid out on the ground as triangles. The many megalithic stone circles of different types, widespread in mainland Britain, most probably had such yet-to-be-discovered features for counting procedures as well as the more familiar alignments to solar and lunar extremes.

Another insight is the number of different units present within the same measurements within Seascale, especially the 102/204-foot measurements. The English yard of 36 inches compares with the shu.si of 0.36 inches so that the yard is 100 shu.si long, and this gives 68 yards (204 feet) in 6800 shu.si, which also equal thirty units of 6.8 feet, the megalithic rod. Also, 2/5 of that rod, the MY of 2.72 feet, is not just a median value within Thom's work but was, it seems, an actual megalithic standard involving the prime number 17 as essential for commensuration of the 6800 days of the nodal period.

The megalithic phase of metrology appears to have been carried forward in the first millennium BCE, when northern tribes mixed with the matriarchal peoples of Greece after the Bronze Age collapse in 1200 BCE. Their temple designs became rectangular cella, which were again able to replicate geometric, geodetic, and harmonic models.

The Megalithic Origins of Greek Temple Design

Temple design in ancient Greece in the ninth century BCE[6] used small dedicated buildings as houses for the gods, then extended these to create, in the Heraion of Samos, a prototype for the hekatompedon ("hundred footer"), whose only entrance faced eastward.[7] The axis was 14 degrees north of east, a familiar angle of a four-square rectangle's diagonal, whose diagonal length stands in relation to that of the four-square base as the solar year

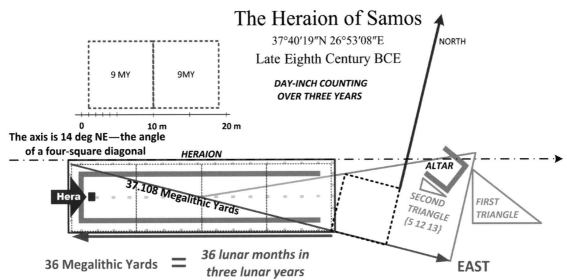

FIGURE 5.9. The Heraion, the temple to Hera on Samos (*top*); and an interpretation of the cella of the Peristyle Heraion (*below*).

Adapted from Hurwit, Art and Culture of Ancient Greece, *figure 33.*

relative to the lunar year. At the western end would have stood a statue, on a podium, of the wife of Zeus, the goddess Hera after whom the monument was dedicated.

Das Heraion von Samos, by H. Walter, showed the monument's early evolution in a scaled plan, and my adapted figure 5.9 indicates a four-square pediment of length 97.96 feet, which would then equal 36 MYs of 2.72 feet (97.92 feet) along the axis. The diagonal is then 37.1 MYs long, these two lengths showing exactly the sort of count found at the Le Manio Quadrilateral in southern Brittany (fig. 1.2, p. 14).

An Early Function for the Peristyle

Early Greek temples had no portico of columns, but in the mid-eighth century, outer "wooden columns were added on stone bases around the long room," says professor of classics Jeffrey Hurwit. He believes this was probably the earliest tentative use of the portico idea but that its lack of utility (being neither structurally necessary nor functionally significant) indicates that it was found aesthetically helpful. "It elaborated, dignified and identified the goddess's house, and from now on the peristyle would clearly and immediately distinguish divine architecture from human."[8] And from it, the notion of the cella, or inner chamber, developed, as seen in Athena's Parthenon.

However, in looking at the composition of the pillars (added to the central ones already required to suspend the roof), it is possible to use them to count two significant astronomical numbers, the first being the number of eclipse seasons (38) in the Saros period of 19 eclipse years (fig. 5.10, *top*). If only the north, west, and south pillars were counted as eclipse seasons, then these outer columns would represent the Saros of 19 eclipse years of 346.62 days. The temple was not merely aesthetic in character but also able to be used to synchronize the ritual calendar with the planetary time periods.[9] And if this is compared to the possible counting scheme, of bays, in Chartres cathedral (c. 1220) in northern France, one sees a continuity of the Greek temple over the intervening two thousand years.

First, 68, the Moon's nodal period divided by 100 days, and 38, the number of eclipse seasons in the Saros eclipse period, show that if Hera was associated with the Moon, then her nodal period would pass 68 columns, each gap representing 100 days, and Hera might represent the lunar maximum or minimum standstills.

HERAION: Counting the 38 Eclipse Seasons in the Saros

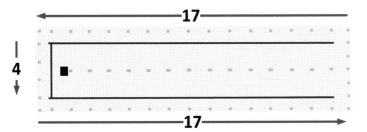

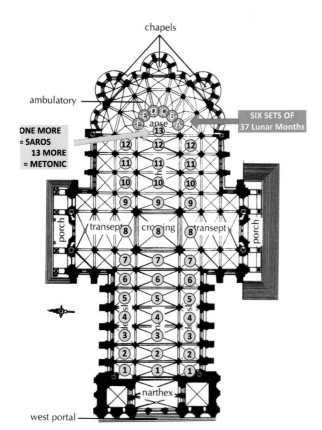

FIGURE 5.10. Comparing the Heraion (*above*) and Chartres Cathedral (*left*). The possible use of the peristyle to count long astronomical periods using lunar months appears translated, two millennia later, in the Gothic period, as the counting of bays—to embrace the counting of Easter using both 12 and 13 month lunar years also found in the solar-lunar calendar of Archaic Greece.

CHARTRES: Counting the Saros and Metonic

The rectalinear form of the Gothic cathedral, punctuated with
pillars, directly extended the form of the early Greek temples,
Greek being one of the meanings of "Gothic." The ability of
number symbolism and metrology to express cosmic time periods
seems to have continued as a driver for the builders of Chartres.

12 + 12 + 13 = 37 lunar months
6 × 37 = 222 lunar months
The Saros = 223 lunar months
The Metonic = 235 lunar months
The difference is 12 lunar months, the Lunar Year

The Transformation of the Cella

Since the Heraion is the first peristyle temple we know of, it must represent why sacred rooms were elongated to form the characteristic cella for the god, the hundred-footer, surrounded soon after by wooden columns to form the peristyle, necessitating an expansion of the platform, initially with stone footings. The first Heraion cella, if identical with the one presented by H. Walter in *Das Heraion von Samos,* would have been a five-square rectangle, as in figure 5.11. But instead, the original cella was a four-square and the new pediment became the same four-square, allowing accurate counting using smaller units (shu.si) on a paved surface. The cella had an extra square added to make it then 32 lunar months long (in MY) while it was the new pediment that was 36 lunar months long—a count over three years. That is, the pediment length became 9/8 the cella's length. The original cella would then be 36 steps of standard canonical Roman feet, long before the Romans of course, since the Drusian module (root 27/25 feet) is 9/8 of the Roman module (root 24/25 feet)

We are then drawn to the conclusion that Hera represented the lunar minimum. Symbolically the outer and inner pillars count up to give 68 pillars, which times 100, gives the days in the nodal cycle. The now-extended cella has a central pillar upon which the light of moonrise would always shine into the cella at minimum, but at no other part of the cycle is that true. The small altar is symmetrically facing the lunar minimum moonset to the north so that cella and altar align to the start and end of the moon's journey through the sky at minimum standstill.

But a more accurate counting of the days of the nodal cycle was also possible, using the flat stone pediment as a counting surface. We have learned through interpreting Seascale that the megalithic yard of 2.72 feet can be used to count the nodal period in shu.si. This type of digit was employed to subdivide the Assyrian foot, but its origins seem likely to have been in counting the nodal period because of the key lengths 102 and 204 feet: the latter is 68 yards and the yard of 3 feet contains 100 shu.si. It turns out that 37.5 megalithic yards (17 rods) contains 3400 shu.si so that 36 megalithic yards plus an extra 1.5 counts 3400 shu.si and the jambs of the cella entrance are inset by a pillar 1.5 megalithic yards from the corner (fig. 5.11). Both sides of the monument can therefore count 6800 shu.si. In lunar months, 37.5 lunar months is three times 12.5 lunar months, which is the synodic period of the ice giant

Hera "as" Lunar Minimum

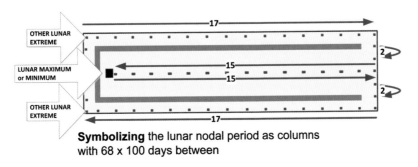

Symbolizing the lunar nodal period as columns
with 68 × 100 days between

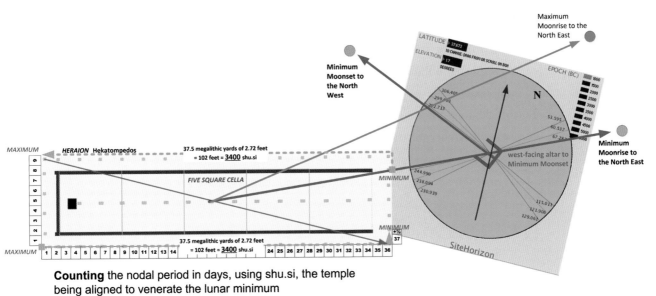

Counting the nodal period in days, using shu.si, the temple
being aligned to venerate the lunar minimum

FIGURE 5.11. Symbolism and counting between Lunar minima at the Heraion.
The cella was extended to being a five-square rectangle and the pediment pillars
became a four-square rectangle. The cella was then aligned with the minimum
moonrise as was the altar to its setting.

beyond Saturn, Uranus. In turn, 12.5 lunar months is 25/24 of the lunar year—an interval called the chromatic semitone because it becomes active when there is a 12-tone scale.

In the dark or Archaic period of Greece, after the late Bronze Age collapse (c. 1200 BCE), the new tribes (whose names came to describe the musical modal scales) entered mainland Greece and the Aegean Islands from northern Europe, where these geometrical forms of knowledge seem to have originated in the megalithic form of the late Stone Age. Samos is an island just west of the adopted homeland of the Ionians, from which tribe the Homeric tradition largely sprang, and so the forms of knowledge seen in the Heraion could be Ionian. The only real difference between the presentation of the second Heraion and the Le Manio Quadrilateral (c. 4000 BCE) of the interval ratio between the solar and lunar year is the now-familiar movement from counting time periods in days to counting then in lunar months* and the transition in the units of counting from inches to MYs, described in chapter 1 (p. 15). In Le Manio, thirty-six stones traverse an accurate 1063 day-inches, while in the Heraion, these symbolic stones have become 36 MYs.

The Heraion cella seems to have evolved from the chambered tombs of the megalithic, aligned in various ways to horizon events marking celestial time, and the peristyle then merged the role of the cairn or mound surrounding the chambered tomb and the stone circle whose circumference was used to mark events on the horizon and whose stones often held counted lengths between horizon events.

LOCATING THE FEATHERED SERPENT

Though historically restricting ourselves to the two easily visible outer planets, the synod of Uranus, the third outer planet, is 396.66 days, which is 1000 of the

---

*As stated in chapter 1, we now know why counting months in MYs usefully displaced day-inch counting, since this normalized the year's excess of solar over lunar years as then being one 12-inch foot (called English), on which historical metrology became based. Normalization happens because the yearly excess is 7/19 (0.368) lunar months, while the MY is 19/7 (2.718) feet long, so that recalibrating day-inch counting to counting lunar months using MYs generates a yearly excess of 7/19 × 19/7, the cancelling of which equals 1 foot. This appears to have initiated metrology as we know it, since 19/7 feet minus 1 foot equals 12/7 feet, the Royal cubit. (For the later British situation, see Robin Heath, *Sun, Moon and Stonehenge,* 81; and for the transition to that, see my own *Sacred Number and the Lords of Time,* 128.)

matrix units (figs. 5.2 and 5.5), which, in turn are 1/80th of the lunar month in days (0.36813 days) to 0.14 percent. This gives Uranus an interval ratio to the Moon of 125/128, so that the Moon is, in relation to Uranus, 128/125 (in the world of Adam's octave of 1440, fig. 5.5). This ratio is the Byzantine foot of 1.024 relative to the English foot, which may be associated with a disharmony this ratio can *eliminate* in relation to what the Egyptians called the Flying Serpent and the Olmec and Maya called the Feathered Serpent Quetzalcoatl (see "The Numbers of Salvation" in chapter 6, p. 133).

Uranus, whether seen or not, starts a new creation raised up by three powers of five, and it is the cornerstone for the Feathered Serpent, just as Saturn is then, 128/125 below Uranus, an interval ratio called the minor diesis. The body of the serpent is a further power of 5 above Uranus, in the next row (fig. 5.12).

In figure 5.12, the upper serpent is shown (shaded blue) in its own right as a recapitulation of the lower serpent. Incidentally, the word *serpent* naturally refers to the serpentine nature of tuning by fifths when populating an octave and may also refer to the serpent of the Garden of Eden story, in Genesis 2, then the serpent on the ground. The main body of the flying variety has its tail above Uranus, in the synod of Mercury (115.88 days) followed by the eclipse year which is 3 times that (346.62 days). In figure 5.5, Adam's matrix for 1440 can "see" Mercury as 1250, above Uranus (1000 units) by the major third of 5/4 (1.25).

Eclipses cannot be harmonically related (using integers) until Adam is

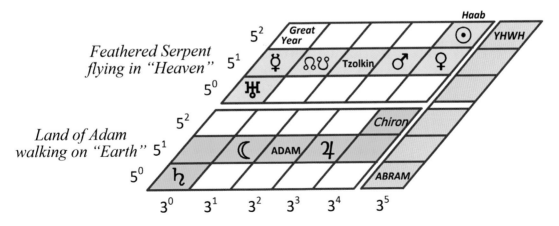

Figure 5.12. Matrix map of the heavenly world of flying and earthly serpents with YHWH as the limiting deity.

doubled once more to 2880, a discovery noted by the Olmec harmonists at Chalcatzingo, Mexico (fig. 5.13). The lunar year is therefore 128/125 below the eclipse year, and a further fifth above reaches the Tzolkin of the Olmec (260 days), now above Adam's head, so that the 18-lunar-month period (of the supplementary glyphs) is 128/125 of two Tzolkin (520 days), seen monumentally at Teotihuacan, as the side length of the Pyramid of the Sun (see also "The Parthenon as a Musical Instrument," later in this chapter). The synod of Mars (780 days) comes next, above Jupiter's 398.88 days, times 125/128 and then doubled to fit the octave (779.3 days).

The name Quetzalcoatl had a triple meaning for Mexico: (a) the Teacher

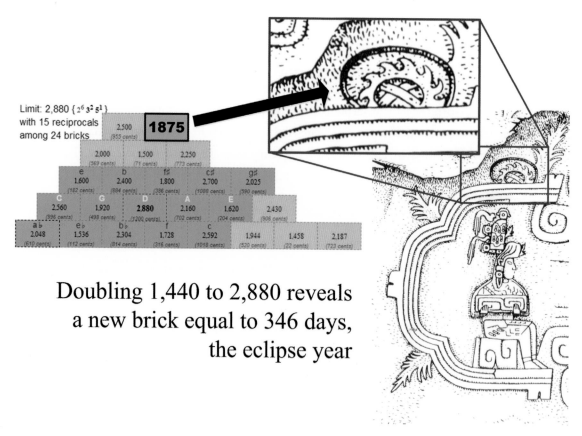

FIGURE 5.13. La Reina ("the Queen"), showing the discovery of the eclipse cycle upon doubling Adam (1440) to 2880.

Drawing of La Reina by Michael D. Coe (1929–2019), from The Olmec World *by Ignacio Bernal, fig. 21 (Berkeley: California U.P. 1969).*
Also featured in Heath (2018), fig. 8.6.

who brought the ancient arts to the Olmec, (b) the Feathered Serpent itself as a whole, and (c) the planet at its head, Venus, brightest planet of all. It is Venus who forms the head of the serpent, with the Haab year of 365 days above the Venus synod (583.92 days) as 5 to 8, the major third once rescaled (to fit a single octave), the common unit being 73 days. It is the year of 365 days that relates also to YHWH, the harmonic limit of geocentric time, as 3/2 of the Haab.

The matrix number of Venus is Plato's sovereign number, 60 to the fourth power (12,960,000), while that of YHWH is 60 to the fifth power (777,600,000), and YHWH flies above Abram the Patriarch (in gematria 243 or 3^5), who in fact represents the sidereal month of 30 sidereal days, 81/80 of the lunar year. Readers can find a fuller exposition of these relationships, related to musical theory, in my book *The Harmonic Origins of the World,* especially chapter 8, on the Olmec, and chapter 9, on YHWH's matrix.

Once the harmonic model expanded numerically to locate the inner planets, eclipse period, and Tzolkin in this way, the higher serpent could become associated with the Second Triangle {5 12 13} because the 13 length could divide into some of the Flying Serpent's time periods while the 12 side could hold the 18-lunar-month period in its base of 12, representing the 12 lunar months in the lunar year. It was noted that the cornerstone of the lower serpent, Saturn, defined the 7-day week since many periods of that lower serpent divide by that week—the Jupiter synod being 57 weeks long (399 days) while Saturn's is 54 weeks long (378 days). With Uranus as the cornerstone of the Flying Serpent the case was somewhat different, and the natural week is 13 days, which divides into the Tzolkin of 260 days as twenty such periods, Mercury as 9 (117 days), and the three eclipse years as 80 (1040 days).

It is obviously the case that the 13-day week of pre-Columbian Mexico, the Tzolkin of 260 days, and the triple signification of Quetzalcoatl survived, while the Old World appears either not to have known of them, or they were kept a secret or even deleted from the Old World through library destruction.

HARMONIC CITIES OF THE SECOND TRIANGLE

The Second Triangle (see p. 37) may have been a hidden esoteric home for knowledge of both the harmonic serpents in unsuspected ways.

Jerusalem

The city of Jerusalem was founded in the middle of the second millennium by the Egyptian Middle Kingdom. Its earliest streets appear aligned to Thurban, the departing pole star, while its more recent "old city" forms a rectangle of 12 units by 5 units so that its diagonals are length 13, and these can then make the Second Triangle (fig. 5.14).

The synods of the outer planets are governed by the cornerstone, which is Saturn, whose natural unitary reality is the 7-day week. The Saturn synod is 54 such weeks, while the Jupiter synod is just 3 weeks longer at 57 weeks. The Saturnian year was 52 weeks (364 days), making the practical year of 365 days, the "year and a day" that was the reign of "matriarchal" kings, men who were sacrificed at the end of their tenure and another chosen, this to respect and *restrict* male power until Jupiter-Zeus came along to upset the reign of Saturn-Cronos on Crete, ending the Bronze Age and initiating the age of iron weapons. It is for this reason, ultimately harmonic, that the outer planets can be modeled on the 12 side by allocating 18 months to 1728 cubits, 96 cubits therefore to a single lunar month, but also, Adam's ultimate development of 1440 would divide 1728 by 1440 to create 6/5 cubits (× 1.2 = 5/4 feet) per matrix unit. This can be seen in that the 5 side of Jerusalem's old city was 720 cubits because that side is 5/12 of the longer side.

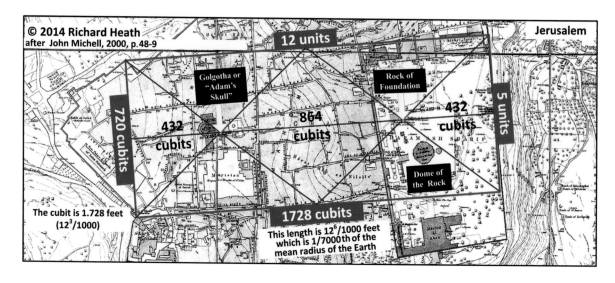

FIGURE 5.14. Geometry and metrology of Jerusalem's Old City.

The Flying Serpent of the Mercury synod, the eclipse year, the Tzolkin, and the Mars synod are all divisible by 13 days instead of 7 days.[10] The cornerstone for this upper register is the subliminal planet Uranus, whose synod of 369.66 days is 1000 matrix units so that the matrix unit is 0.369132 days, which is numerically similar as one-thousandth of it. This makes the 13 side a perfect length into which a 13-day week of day numbers, as used by the Maya, will divide. Primarily, it is the eclipse year, the Tzolkin, and Mars that are meaningful. Three eclipse periods add up to forty 13-day weeks or 520 days (40 × 13), which is two Tzolkin of 260 days. This was geometrically modeled at Teotihuacan (and see also "The Parthenon as a Musical Instrument," later in this chapter), and it demonstrates the movement in powers of three leading to the musical interval of a fifth (ratio 3/2). Note that the cubit (3/2 feet) can naturally divide up a counted length through using its reciprocal of 2/3 feet.

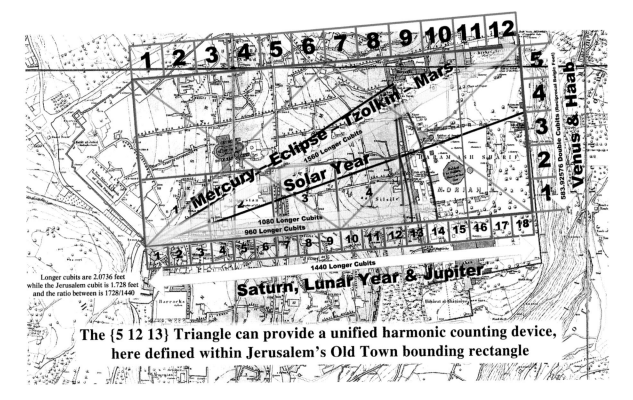

FIGURE 5.15. The use of the Second Rectangle to map out both giant planets on the base and inner planets on the diagonal. The 5 side can be used to map Venus and the Haab of 365 days.

The parts of the Flying Serpent sit above the tones of the outer planets by the minor diesis of 125/128, a ratio that multiplies by the third power of 5 (125) while clearing seven powers of 2 (128). Musically, this interval is the sum of three major thirds (5/4), which will then fall short of a full octave doubling. Thus, the inner planets are *disharmonized* with respect to the outer planetary world of Adam, and myths such as those of arks and of New Jerusalem are likely metaphors for moving humanity into harmony with the inner planets through which their microcosmic soul can be formed, rather than the outer planets, which are responsible for the macrocosmic, sublunary world. This could explain why the inner life of humans is, in the absence of such harmonization, dominated by the law of accident—that is, people not knowing what they actually want.

The design of Jerusalem appears to be a natural usage of the Second Rectangle within cities and monuments, as with the cella of the Parthenon (chapter 1). Jerusalem, unlike the Parthenon (see the next section), holds the fuller harmonic model of a "serpent on the ground" (outer planets) and a "Flying Serpent" (inner planets). The numbers 3 (a factor in 12) and 5 are in two of its sides, while 13 (associated with eclipses and the 13-month lunar year) divides into time counts such as the Mars synod of 60 weeks of 13 days (780 days).

The 5 side has its own attractions for modeling Venus since the 584-day synod is 8/5 (1.6) of the practical year of 365 days, this also the Mayan Haab. The common factor is 73 days long, with 8 in the Venus synod and 5 in the year. The axis of the city was associated with the entrance toward the east indicating Shekinah, the goddess of the land of Israel.

The Parthenon as a Musical Instrument

To *create* the harmonic model of sky events, the synodic periods had to be quantified by the megalithic astronomers using triangles and noting the time counts between the loops of the outer planets (see "The Harmonic Planetary Model" section earlier). The counts are then seen like musical string lengths, which led to the instrument model being used to express them, for example at Teotihuacan. One of the clearest is within the Parthenon, couched within the Second Triangle, which is already strongly linked to the Lunation Triangle (see the discussion of the Heraion on p. 115). (It was the Parthenon that first allowed me to find the harmonic model within a monument.)

The Parthenon is a rectangular monument whose longer dimensions are used like strings of different tone, according to musical intervals (fig. 5.16). However, the vibrations are then of the days counted, the longer the count the higher the pitch, opposite to a musical instrument, where a strings pitch goes down as musical strings get longer. In other words, the harmonic model is based on each day as a single cycle, and the cycles accumulate to form counts of higher frequency.

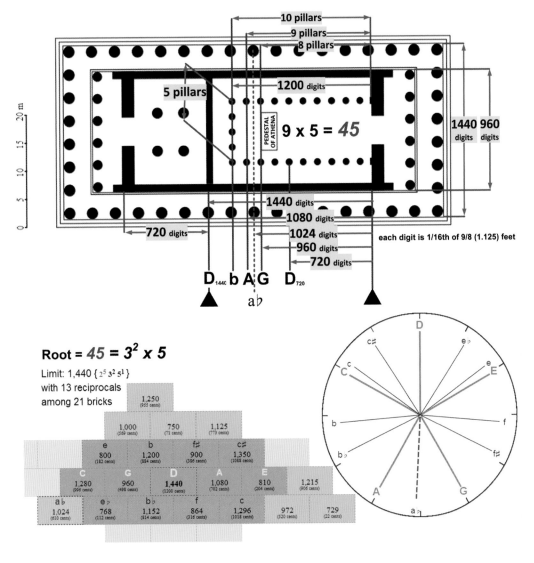

FIGURE 5.16. The Parthenon as a musical instrument model of the Moon and the inner planets.

I call such musical models a matrix, and here the matrix unit is represented by a measure called a digit which represents 1/80th of the lunar orbit. The basic scale then is 80 digits to a lunar month so that 12 lunar months, the lunar year, make up a string length of 960 digits. But the octave (which these models imply) is of 18 lunar months (1440 digits), the periodicity used by the Maya for their supplementary glyphs to some Long Counts. This brings us to the conclusion that this octave and its models were esoteric, even though represented in great monuments and also in some texts, since the harmonic root is 45 (1440 is 32×45) and in the Bible Adam's number is 45 but also (using position notation) 1440. Here Adam is Athena whose temple this is, and she is a transformed type of Moon goddess, and the harmonic model is all about how the Moon connects the Earth to the outer planets. Athena came out of Zeus/Jupiter's head after he ate the matriarchal goddess Metis, and in this model, Jupiter equals 1080 matrix units, a number associated with the Moon, because Jupiter controls the Moon with his 9/8 whole tone between 1080 and 960 units.

Teotihuacan

This Mexican city was the first and largest ever built in the pre-Columbian Americas. It emerged around 200 BCE, having evolved from the founding Olmec culture of southern Mexico. The Olmec first appeared, at the end of the Atlantic "conveyor" of trade winds and ocean currents, around the same date as the foundation of Jerusalem. What is unique about the city is its sacred precinct with three large pyramids, one of which has an important association to Quetzalcoatl. The central axis points east of north, also to Thurban, and it has a long "road" down the longest dimension, linking the whole site but also allowing long counts to be quantified.

Its architectural units appear to be the megalithic yard, called the Teotihuacan measurement unit (TMU),* a unit of measure that, in feet, resembles the reciprocal of the matrix unit, but then in days. The matrix unit is the reciprocal of 2.709, and the nearest MY measurement, from megalithic times, is the true astronomical ratio of Sun and Moon, 2.715, which is almost identical. Archaeology professor Saburo Sugiyama found astronomical numbers sacred to

---

*After twenty years of study, Sugiyama found good correlations within the monument to the MY, which he called the Teotihuacan measurement unit.

the Olmec in the Sun Pyramid (260 TMUs), Moon Pyramid (173.3 TMUs), and Courtyard of Quetzalcoatl (584 TMUs), explicitly in these units. These are the Tzolkin, half of the eclipse year, and the Venus synod, respectively. He also located the distance between the center of the Sun Pyramid to the center of the Quetzalcoatl Pyramid as 1440 TMUs, which is correct for Adam's matrix of time based on the Moon and the outer planets.

It therefore appears that the TMU equaling 1 day was employed when signifying the Feathered Serpent, while the road and the length between pyramids, of 1440 TMUs, was coded in matrix units. This would have provided a means to integrating both the serpents within a site with very large buildings that were connected by a road that could be used to measure lengths with long ropes and then permit ceremonial processions as an outer connecting function.

We can detect the Second Triangle in the key width and length of the complex. The width is 1040 TMUs, which is four times the Tzolkin (260 days), while the length given by Sugiyama is 2600 TMUs.[11] Dividing 2600 by 1040 gives 5/2 (2.5), which is 25/24 of the Second Rectangle (2.4), and so it is very close to that triangle and may have been intended so, using slightly different datum points than the professor did. In figure 5.17, the matrix model runs from the Pyramid of the Sun to the south (*right,* as shown), and the key lengths of the lunar year (the Saturn and Jupiter synods) bridge the San Juan canal. The Venus synod would be 1582 matrix units, hence TMUs, which accurately delineates the Temple of Quetzalcoatl.

To the left of the harmonic right triangle seen in figure 15.17, an amazing geometrical fact relating to the 3/2 ratio is shown. The 3/2 ratio is the musical fifth (and the metrological cubit), here shown as a 3 side of 780 TMUs (Mars synod in days) running along the road, and a 2 side at right angles running 520 TMUs (two Tzolkin in days), to the top side of the overall rectangle in figure 5.17. Miracle of miracles, though it can be proven true through analysis, the third side between the Sun Pyramid and the platform before the Moon Pyramid is 937.5 TMUs, which, *in matrix units,* corresponds to the eclipse year (346 days).

These three periods (eclipse year, Tzolkin, and Martian synod) are adjacent within the Feathered Serpent, except that the triangle with sides 2 and 3 has automatically calculated the third time period in matrix units. This is symbolism, knowledge, and brevity of a high order, and the 937.5-TMU length could be taken, as a rope, and placed into the matrix model to the right of the solar pyramid to locate where the eclipse year would be in the matrix, were it an integer. And to make it integer, the 1440 limit of Adam would need to be doubled to 2880, whereupon its value would be 1875 (as per fig. 5.13, p. 122). Doubling is easily achieved by using half TMUs.

The Moon's nodal period can also be located in the matrix of Adam as 1440, and 6800 days divided by matrix units reduces down to the octave for 1440 as a non-integer that, rounded up, is then 1152 which is located on the bottom row, third brick (note b-flat in fig. 6.2, p. 136). This implies the nodal period of 18.618 years has a relationship to the Saturn synod of 9 to 8 (9/8 or whole tone). 378 days times 9 divided by 8 gives a time period one-sixteenth of 6800 days. Readers can also move forward to the section on the Canterbury pavement ("Canterbury's Cosmic Conundrum," p. 172) to see Saturn correlated with the nodal cycle.

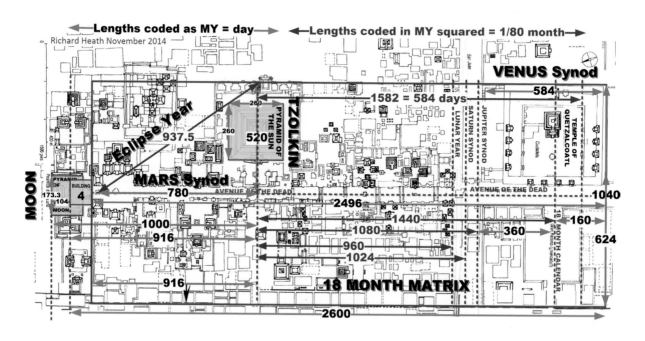

FIGURE 5.17. The sacred precinct of the Mayan city of Teotihuacan was doubly coded, representing the cosmic serpent as a musical matrix for 1440 (*right* of the Sun Pyramid) and a harmonic right triangle with sides 3 and 2 (*left* of that pyramid) in MYs.

6

The Savior God

In Christianity, the oneness of God is presented as his Son. Jesus Christ embodied the idea that each human is a child of God. Such a son of God had been prefigured in Old Testament stories such as patriarch Abraham's near sacrifice of Isaac, then amplified by later prophets into a Jewish passion for a messiah or savior. This provided the context for the life of Jesus.

The oneness of creation takes the form of the Mother, who is three-fold as Sky, Earth, and Moon. Oneness was also felt to originate in the first three numbers: 1, 2, and 3, which form the creative matrix for what follows as the later cardinal numbers. This triad of 1-2-3, standing atop the number pyramids of the Pythagoreans, identifies the number 2 as female (as did Plato) since it is reproducing through an evenness that is barren without the oddness of the male number 3 that can populate that womb with musical scales, that is, with notes. This pre-classical idea stands opposed to our habituated idea that numbers merely represent objects and measurements rather than symbolizing active principles. But you can go too far with number symbolism, and its practical purpose as framing the creative patterns that were precursors to this world and prior to the time and space of existence can be missed.

J. G Bennett suggested there were four major views of God: as Father, Mother, Sacrificial Son, and Spirit.[1] These, he thought, must have developed in regions isolated from each other, involving the creation of language, and those stories epitomizing these different views.

1. Dynastic Egypt was Father-oriented,
2. The Mediterranean was Mother-oriented—with a matriarchal organization of society,
3. The savior god was seen in the myths of the Indo-Europeans, and
4. Contacts with the East and the New World presented religious ideas centered on the Great Spirit.

In 1965, Bennett proposed these four views had largely been achieved in regional centers, in Europe, in southwest Asia, in Africa, and in the Far East. Populations in each of these regions were relatively *unaware* of these objectives while their leaders were often *psychotelios,* that is, psychoactive initiates to angelic sciences and to the *sacredness* each image of God contained. Developing in relative isolation, these four types of divine association with father, mother, child, or spirit world later became mixed up through the later historical processes of climate change and the ages of stone, bronze, iron, and technology.

Abraham's Hebrew story is set toward the end of the Bronze Age, an age in which only the god-kings, heroes, priests, and officers of state were counted as significant actors. This age would collapse suddenly, in the few years around 1200 BCE, a collapse that freed up the Archaic Greeks from surrounding empires. They were inheritors of what are now called Greek myths, and throughout their "dark age," this enormous oral corpus of myths was transformed as the patriarchal Indo-European tribes interpenetrated the matriarchal peoples of Greece and their goddesses. A growing individualism of thought and action gave rise to the first significantly literate population as the Greek phonetic alphabet gave freedom from the cultural domination that accompanies the recitation of oral stories, as sole touchstone and mechanism for collective memory.

The above sketch of a developmental history for humanity can stand apart from *history-making* histories, because human development is not wholly dominated by recorded facts. In the same way that life builds its capacities through interaction with environments, there are significant archetypes at work, such as the Greek sense of *individuality* and selfhood, this deriving from the oneness of the person and the subjective nature of selfhood itself. We can feel the dynamics between father, mother, and son or daughter, in the birth of a human, as the four above types of religious idealism, in one sentence, as a new human spirit enters the world. This symbolism in the working of sacred images came to decorate the religious spaces of Christianity, like geometrical models and metrological ratios, but especially focused on the microcosmic truth. Beneath Christian images, a developmental work on sensibilities was patiently operating.

It was essential that the savior Jesus be Jewish because of the contacts the Hebrews had made with Egypt, the Mediterranean, Zoroastrianism, and so on.[2] Their early Bible was already a unique synthesis of ancient stories, metrology, and musical octave theory, with their history written while living in the Babylonian exile. There had been many phases: patriarchs, judges, kings, and prophets that foretold something about a future time we call destiny, or future history.

THE NUMBERS OF SALVATION

The esoteric traditions of the ancient world were swept up into each new Abrahamic religion, and this led to an art and architecture that expresses the necessary human salvation from a cosmos and world created by the Universal Will, in which the human can be saved and have a higher destiny than that of the fate of ordinary life. There is every manner of wrongness possible in how this new relationship with God might be motivated, and so a purification of intent is involved as is faith that there is a God and that there can be more than ordinary life as its purpose.

The Minor Diesis of Byzantium
In the harmonic model of chapter 5, we find the Feathered Serpent of inner planetary synods, the eclipse period, and the Tzolkin all raised by three major thirds (125/128), making them disharmonious to the lunar year and outer planetary synods. This separation of inner and outer planets, by 125/128, suggests the octave model was like the holy mountains conceived in the angelic design, though probably not requiring the notion of a mountain, but rather a direct perception of the angel in its world. The Byzantine foot of 1.024 feet (128/125)*

---

*That is, the reciprocal of the 125/128 ratio called a minor diesis or $5^3/2^8$ divided by 2 to fit a single octave.

enables the lunar year of 354.367 feet to become 346.0615234, the eclipse year to one part in 600. Therefore, this foot provides the required linkage between the inner and outer planetary registers.

The Byzantine foot of 1.024 feet (128/125) could represent what was required for reunification of the humans on Earth with their true nature (microcosm), which is represented by the inner (or personal) planets, in a kind of New Jerusalem with a new song sung by choir of 144,000, raised by a minor diesis above the bottom row of the harmonic matrix, then sitting beneath the Feathered Serpent and above Adam as the octave 9 to 18 lunar months. The New Jerusalem was a type of city as ark, and the ancient Near-Eastern notion of an ark involved the floating upward to, most classically, three powers of five (in harmonic terms), to the limiting number of 1,728,000, which is $12^3 \times 10^3$. If 1000 (10^3) is taken to be the three dimensions of a cube, then,

FIGURE 6.1. The Deisis* at the Hagia Sophia, in which Jesus has John the Baptist to his left and the Virgin Mary to his right.

Retouched photo by Myrabella for Wikimedia Commons.

---

*Deesis (or Deisis) seems uncannily like the musical semitone of the Pythagoreans 128/125, the minor diesis. Both derive from Greek. Deisis is the Greek word δέησις, meaning "prayer," or "supplication," while in Ancient Greek δίεσις or díesis means "sending through, the smallest interval in the scale." It is improvably possible that prayer and semitone were once esoterically equated.

in an act of primitive numeracy, the head number of New Jerusalem is 12 to the power of 3, 12 × 12 × 12 = 1728, which equals 1000 times the Jerusalem foot of 1.728 feet, 1,728,000 being the number of most textual arks predating the Bible, whose design was expressed as 1728,000.* But now we can see that these numbers are matrix numbers, keyed to one-eightieth of the lunar month. In 1728,000, if the prime number 2 is removed (as the cornerstone of the matrix $2^9 = 512$ in this case), the result is 3375, which is $3^3 × 5^3$ (that is, 27 × 125), the location of the ark. This leads to a new interpretation of Christian salvation, the city as an ark as described in The Book of Revelation illustrated in figure 6.2. Further background on this can be found in my book *The Harmonic Origins of the World.*

The twelve tribes of the New Jerusalem are in minor diesis (125/128) to Saturn, and the Moon's nodes are in minor diesis to the choir of 144,000 singers, who sing a new song, a song based on the inner experience in harmony with outer experience, through the minor diesis of the Byzantine foot, as Christ who has redeemed them and raised them up.

The Essential Duality of Saturn and Jupiter

In the next chapter, you will see how it becomes possible to draw together the Domes of the Hagia Sophia, the Dome of the Rock, and the Christian pavements of Britain (especially that of Canterbury) as having the model of equal area that is of the Moon's nodes. For example, the presentation of Saturn as a diamond (and a tritone) relative to the 9:18 lunar month octave has an exact relationship to the lunar nodal period of 6800 days, which is 18 times 378.09 days of the synod of Saturn. One can divide 6800 days by 378 to get 18 synods as its length, within five and one-half days. This places the nodal cycle as 9/8 if transposed to Adam's matrix, Adam now being transformed into Jesus and dying on Golgotha, meaning "Adam's skull" (shown in fig. 5.4, p. 124). Saturn is a♭, and the nodal cycle is "down on the Earth" with the nodes as b♭♭, while five Saturn synods equals 64 lunar months on the second row of the matrix.

In the bigger picture of figure 6.2, Saturn is the first note—a tritone to

---

*In numbers such as this, the nonzero numbers at the start of a decimal number are called its head number because it is made up of prime number factors {2 3 5}, while the tail of zeros is made up of only {2 5}. Any comma notation in head numbers is dropped when this is a factor since such head numbers are not decimal and have a nondecimal numerical significance.

Christ—in the sense of Life on Earth and also the cornerstone of the matrix and hence the definer of it, while the nodes are a whole tone to the Saturn synod, just as the Jupiter synod is a whole tone to the lunar year in the row above. These two rows represent human fate and, in the Canterbury pavement, the diamond is juxtaposed with the nodal period through the junctions with it of the four circles to form a cross.* This harmonic view of the nodal cycles is congruent with traditional stories going back millennia, such as those discussed in *Hamlet's Mill,* referring to the mill wheel or quern used to express the grinding nature of the movements of the stars and planets as the top stone and the Earth as the bottom stone, with life's experiences between, over long ages of development. In that tradition, which is quite uniform yet various in the creative metaphors deployed to describe the same phenomenon—the precession of

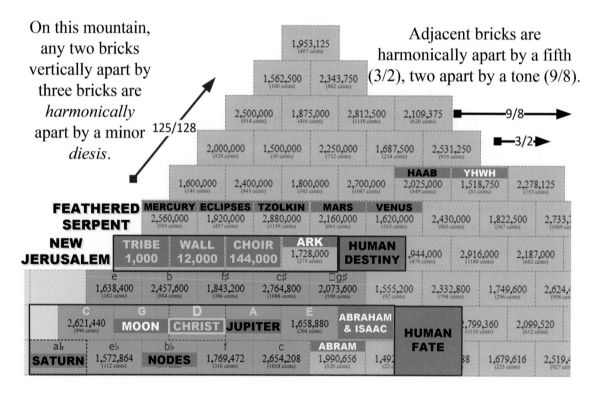

FIGURE 6.2. The proximity of New Jerusalem to the Feathered Serpent
(see also fig. 5.12 on p. 121).

---

*A Maltese style of cross pattée, which includes variants of the red cross of Saint George, who kills the dragon.

the equinoxes—we see Saturn as the maker of the mill that Jupiter (Hamlet's uncle) steals to become king, just as Zeus-Jupiter deposed Cronos-Saturn in Hesiod's *Theogony* of the late eighth century BCE.

The mill is the semitone of 16/15 to Zeus-Jupiter, just as the Moon is the semitone to the cornerstone of Cronos-Saturn. But Jupiter has also seized the Moon with the whole tone of 9/8 to the lunar year (a more powerful resonance) while yet holding the nodal cycle with the semitone (an inferior grip to Saturn's on the nodal cycle). The stronger power of a ruler is accompanied with an inferior grasp of human fate that, when projected into a human life, makes the mill a liability in their hands. Our diminished spiritual solutions to the problems of life and strengthened needs of consumerism are similarly blind to consequences.

Saturn is the legitimate and intended boundary of both the solar system and the view from Earth as the outermost visible planet, traditionally imposing fixity and limits for the factual world of subjective time and space *that is* the microcosmos. Saturn is a necessary denying force, a force defining and maintaining the relative quantification one sees as factually numeric, for existing things must be concrete to then have a quantifiable nature. Saturn is therefore Satan by another name, who makes actualization possible for existence on Earth, a principle complementary to Jupiter, the two being seen as twin dark and light forces (in extremis, the Moon and the Sun), called by some Ahriman (the Zoroastrian god of darkness) and fallen angel Lucifer, both trapped beneath the arch of time while driving all the desires and wishes of the sublunary world. These two forces originated in Persia, in a doctrine of necessary balance between them, expressed within the Neolithic agrarian revolution and as the idea of Christ. The Neolithic period supported the populations and cities that gave us civilization, yet the third force—of reconciliation—cannot be based on greed and other "deadly sins." Jupiter and the god-king became affirming for civilization, and Saturn's denial of possibilities was then seen as restrictive to growth because the three-term system (the triad of three impulses; see "Ninefoldness as Universal Scheme," p. 223) easily collapses into a two-term dichotomy (dyad of two polar positions).[3]

The reconciling force is the lunar month that presents, in its phases of light and dark, both Sun and Moon, in a cycle of illumination by the Sun, the primary source of all illumination and of what can be seen. The Canterbury pavement (fig. 7.21, p. 172) seems to place the Cross within the context of

Saturn's diamond, as the promised Comforter. In the centuries following the
life of Jesus, his image became transformed from the early Christian art of
Rome and later Christian art of Byzantium, from being solar or Jovian to being
Saturnine. That is, Jesus became transformed into the *pantocrator* (or Lord of
the World), and in his *deisis/deesis* imagery he becomes Saturnine in appearance.
This convention began in the sixth century and was "of a far more sweeping
character," becoming "essentially ecclesiastical, instead of Scriptural," according
to the British archaeologist J. Romilly Allen.[4] In other words, the subject of
what figurative Christian art should be, only became resolved and centralized
in Constantinople and not in Rome. The early art of the catacombs expressed
narrative snapshots such as a happy Good Shepherd carrying a lamb (the lost
sheep of repentance), but later religious stories became ensnared in the greater
narratives of guilt, history, and official dogmas that were policed. The *structural
art* of a mosaic, pavement, or a building's dimensions were happier though, left
to the craftsman, the guilds, and their masters.

If Christianity was to liberate a portion of humanity from the excesses of
Jupiter, through the dynamism of the Moon, Jesus must be a king (of the Jews)
and Saturn must become a thorn according to the Sufi adage "Use a thorn
to remove a thorn"—an endless fate for an unspiritualized human race. At
Canterbury, the Cross of Jesus was shown within the diamond of Saturn so as
to take on the sins of the human world. It is fitting then that Thomas Becket's
shrine was installed alongside it since Becket (a) spent many hours contemplat-
ing the pavement from his exclusive throne, overlooking it, and (b) Henry II's
men dashed him to the ground in 1170 for opposing reform of the Church's
powers over the king. The four roundels then indicate Jupiter's synod in compe-
tition with Saturn's, presented as an increasingly Saturnine Jesus from the sixth
century onward.

Once the time-world of the planetary matrix is seen in this way, narratives
become connected to that very real phenomenon. In describing the planetary
world of time and its harmony, the direct relevance of the matrix technique
to religious symbolism becomes clear. The only obstacle is that ecclesiastical
authorities do not need a factual basis for religion. Instead, they want their
respective organizations, like any other, to continue on indefinitely without
any change, certainly on this scale. And before approaching the strange story
of England's sacred pavements, then Islam's Kaaba, there is more to say about

the meaning discoverable within the Seascale stone circle since it appears to be a model of the nodal cycle that is held between Jupiter and Saturn.

Could Prehistory Really Quantify the Nodal Period?

I would suggest that instead of counting the 6800-day nodal cycle in day-feet (it would then be over 1.2 miles long), the length was halved in size, allowing continuous counting, to and fro, from maximum lunar standstill to minimum lunar standstill, over 3400 days,* and then back again. When the height (81.6 feet) of Seascale's four-square rectangle is measured in shu.si (of 0.03 feet), the result is 2720 shu.si, exactly 4/5 of 3400 shu.si. By adding one more side length (by turning left at the top-right corner,) a half-perimeter day-count, between bottom right and top left, was then 3400 shu.si (see the left rectangle of fig. 6.3 on page 140). This is remarkable since, already, the *diagonal* between opposed corners is also the count of the solar year in lunar months as 12.369 megalithic rods. There must be an exact relationship between the solar year and the nodal period that had been quantified within the units (megalithic rod and shu.si) and the four-square geometry. In gravitational physics, the nodal period is caused by gravitational influences on the Moon, from either side of the plane of the Moon's orbit, notably those of the Sun and Jupiter. The metrological and geometrical mindset of the megalithic period had instead applied itself to the ratios between the periodicities of the lunar year, planetary synods, and long-term solar-lunar recurrences within the four-square rectangle.

To obtain a 6800-day shu.si measurement around Seascale's four-square rectangle, 6800 day-feet would have to be divided by 1000 to obtain what Thom called a megalithic rod of 6.8 feet, which was then multiplied by 3 so that the side length of each square was 20.4 feet. There are then 10 side lengths in the rectangle's perimeter and therefore 204 feet (12 × 17) around the whole perimeter. But remember that only a unit (or measurement) with 17 in it can divide into 6800 to give an integer result. The factors of 6800 are 400 × 17, and 12 divided by 400 is 3/100, the exact formula for the shu.si in feet. Similarly, the megalithic rod of 6.8 feet is 17/10, and hence the megalithic rod of 2.72 feet is 68/25 (since 272 is 16 × 17). This reduction can be seen in figure 6.3.

---

*The Aubrey Circle is around 3400 day-inches in diameter, or 283.3 feet, reducing the length of a nodal count.

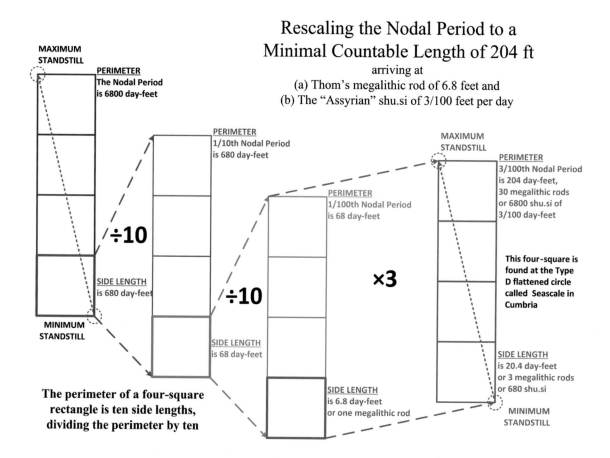

Rescaling the Nodal Period to a Minimal Countable Length of 204 ft

arriving at
(a) Thom's megalithic rod of 6.8 feet and
(b) The "Assyrian" shu.si of 3/100 feet per day

MAXIMUM STANDSTILL

PERIMETER
The Nodal Period
is 6800 day-feet

÷10

SIDE LENGTH
is 680 day-feet

MINIMUM STANDSTILL

The perimeter of a four-square
rectangle is ten side lengths,
dividing the perimeter by ten

PERIMETER
1/10th Nodal Period
is 680 day-feet

÷10

SIDE LENGTH
is 68 day-feet

PERIMETER
1/100th Nodal Period
is 68 day-feet

×3

SIDE LENGTH
is 6.8 day-feet
or one megalithic rod

MAXIMUM STANDSTILL

PERIMETER
3/100th Nodal Period
is 204 day-feet,
30 megalithic rods
or 6800 shu.si of
3/100 day-feet

This four-square is
found at the Type
D flattened circle
called Seascale in
Cumbria

SIDE LENGTH
is 20.4 day-feet
or 3 megalithic rods
or 680 shu.si

MINIMUM STANDSTILL

FIGURE 6.3. Visualization as to how 6800 day-feet
might be reduced in size by 3/100.

In appendix 1, I term Thom's earlier MY *nodal** because (as above) it
must have come into existence through identifying that 6.8 was a factor of
6800 and through using the rubric of the metrological step of 2.5 feet; that
is, 6.8 feet divided by 2.5 being 2.72 feet. However, 2.72 feet is not a foot
but rather is within the range (in feet) of the lengths we call MYs. I believe
the four-square rectangle of Seascale had a perimeter of 30 megalithic rods
of 6.8 feet, which is 204 feet, the length that divides into 6800 parts of

---

*The nodal megalithic yard (or NMY) of 2.72 feet, see also appendix 1 on shu.si and
megalithic yards.

3/100 feet, the shu.si—evidently, also, conceived as a nodal counting length for the day.

It is widely accepted that the megalithic people established long alignments to accurately signify the moment at which the lunar maximum and minimum occurred. Thom was asked, "And what happened if, after 18.618 years of waiting, they missed it?" He dryly replied it was just too bad. However, this inherent tragedy could be, and was, avoided by counting the days in-between such events, which not only readied astronomers for the approach of a singular event but also led to a continuous set of ongoing counts of different cycles. This must have given everyday life to megalithic observatories, in which the interactions between time's cycles could be quantified geometrically and where the lunar year, solar year, nodal cycle, and Jupiter synod were tracked using counting, sometimes around circles representing the Ecliptic path of the Sun, Moon, and planets on which the stars were fixed. It is also true that such observatories would set the time for other observatories: traveling between observatories, it was possible to encode the count, as the Maya did, while counting the days spent traveling between observatories. Judging by the ubiquity of nodal alignments in the megalithic era, the nodal period became a standard longer system of time, almost certainly associated with death and transmigration beliefs.

A further time cycle was easily achieved within Seascale: the synod of Jupiter as 9/8 of the lunar year. The rectangle's length was the 12 months of the lunar year, in megalithic rods, so that the Jupiter synod should be a length 13.5 megalithic rods long. Because the rectangle is four-square, each square can be divided by 2. Exactly halfway along the north side of the rectangle is 13.5 megalithic rods (91.8 feet) from its bottom left. A count from a bottom corner to halfway across its north side is therefore the Jupiter synod in rods per lunar month. This is also 10/9 of the top left corner since $10/9 \times 9/8 = 10/8 = 5/4$: there are four sides and then the fifth, the top side.

The influence of Jupiter on the nodal period is in its proximity, through the 3/100 scaling, with the precession of the lunar orbit, making it 6800 days. Normally, the month is counted using the MY of 2.72 feet, or 32.64 inches, which can be compared to the lunar month as the ratio $32.64/29.53125 = 1.105\mathbf{269}$. Strangely, the synod of Jupiter can be compared to the 361-day sidereal period in which Jupiter passes through a Zodiacal sign as $399/361 = 21/19 = 1.105\mathbf{263}$. These near-identical ratios reflect the fact that Jupiter is an influence on the nodal cycle, based on the Moon being resonant to Jupiter and appearing full in the center of its synodic loop. The lunar month is the synod of the Moon with the Sun, while the Jovian month, when the Moon again catches up with Jupiter, is the synod of the Moon with Jupiter (27.5 days).

The Moon is held between the Sun (traditionally its first ruler and sometimes conflated with Saturn) and Jupiter, who was the stealer of whatever Hamlet's Mill was. In the book *Hamlet's Mill*,[5] it is said that the deviances of orbit of the Moon with respect to the Sun (the nodal period over 6800 days) and of the Sun with respect to the celestial Earth (the precession of the equinoxes over 25,920 years) were conflated in the myth of a grinding mill shaping human fate and destiny. The notion of a sacrificial god, who saves the world by dying on the World Tree, is a parallel vision of human fate and destiny as emanating from these cosmic deviances rather than from the regularities of the machine in the sky. The Abrahamic religions inherited much of this ancient lore as a basis for their symbolism of the divinely created world. Ancient details came into conflict with the proposed finality of scriptural accounts, though beneath their symbolism still lie, like a sleeping dragon, the facts from which such symbols had come into existence. In built structures, metrology lived on and continued to present counted geometrical forms.

The purpose of human life is mocked by atheistic materialism and held back by ecclesiastical obsolescence, apparently ignorant of the actual location of heaven. The framework conditions, which have created the destiny of humans, were thought to extend downward into the human soul as a microcosm that was made in the image and likeness of the macrocosmic Will. As an image, this is the descending, sacrificial god on the World Tree or Woden's cross.

7

Pavements of the Savior

IT IS A PROBLEM TO INTERPRET A MONUMENT without first understanding the design problem it solved. In the previous chapters we resolved some major astronomical and geometrical models influencing the way monuments were built over the period leading into the Christian world. I now offer some interesting examples garnered from the designs for important church pavements in England.

I became interested in the Christian pavements through the Cosmati pavement at Westminster Abbey.[1] Only recently, I saw that this famous coronation pavement conformed to the Equal Perimeter Model of John Michell. I then found that Canterbury Cathedral had an even more enigmatic design, and, in reviewing John Michell's work on Glastonbury Abbey, I encountered an extraordinary state of affairs there: a destroyed pavement only lives on through geometrical inference and "automatic" writing, from a subconscious, spiritual, or supernatural source! And, perhaps, the Westminster pavement design was based, in part, on the first- to fifth-century Glastonbury pavement, and its environs before the Lady Chapel housing it was ruined by Henry VIII during the disestablishment of the monasteries.

Circular Christian pavements were high-status objects once churches became formalized, though few survive. The most famous is Chartres's famous labyrinth in France; many Gothic labyrinths were lost to the French Revolution. At Chartres, the rose window is vertically placed between the western towers, at the same height as the horizontal distance to the paved labyrinth from that wall's footings. Hence, it may be true that for the medieval Gothic designer, circular windows employing colored glass were replacing circular pavements and their semiprecious stones.

An interesting place to start this story of three pavements is at Glastonbury, once an ecclesiastical center for England. The work of John Michell and others has reconstituted some picture of Glastonbury's earliest church and its pavement.

THE STORY AT GLASTONBURY

Glastonbury has an enigmatic history connected to both Arthurian and Early Christian legends, and, before that, early histories point toward Glastonbury as a sacred Celtic center. Its early topography was unusual: an island within a region of shallow water, now reclaimed to form the Somerset Levels. The island had an improbable hill called Glastonbury Tor (fig. 7.1), now surmounted by a Gothic tower dedicated to Saint Michael.

The Christian histories of Glastonbury Abbey are prefixed with its mythic establishment as early as 37 CE, by Joseph of Arimathea,* whose tomb

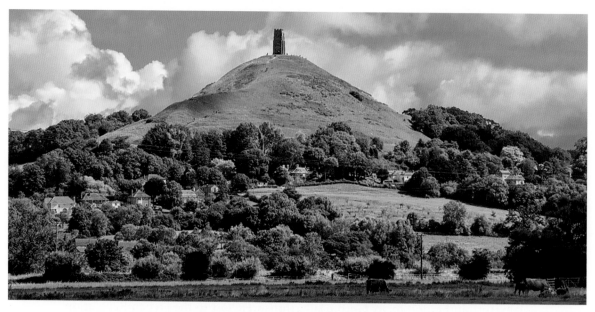

FIGURE 7.1. Glastonbury Tor with St. Michael's Tower.

Photo by Eugene Birchall for Wikimedia Commons.

---

*Saint Gildas, a sixth-century British monk and historian, wrote "Joseph introduced Christianity into Britain in the last year of the reign of Tiberius," that is, 36–37 CE. From *De Excidio et Conquestu Britanniae*.

had to be used for the body of Jesus after crucifixion (fig. 7.2a). A more likely date of establishment is 63 CE, by Saint Phillip of Gaul, but if not that, then certainly by 516 CE in association with historic British saints such as Saint Patrick. By all accounts, a small circular church was built at Glastonbury, made using the local wattle-and-daub technique. Some say this first Christian hermitage was dedicated to the Virgin Mary so that the still-standing Gothic

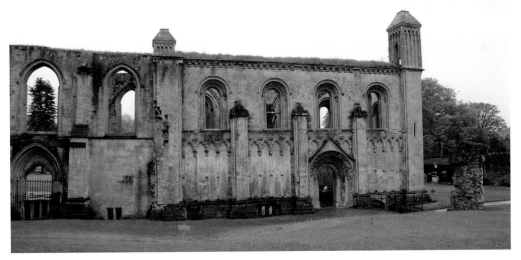

FIGURE 7.2. *Top,* imagined early Christian foundation at Glastonbury.

Illustration by Bligh Bond, permission granted by the Glastonbury Abbey.

Bottom, the Lady Chapel built to house the circular pavement
of the early wattle-and-daub church at the center.

Photo by Karl Gruber for Wikimedia Commons.

building built to protect the pavement, after the old church burned down, was called the Lady Chapel.

As with the earliest Greek rectangular temples, the vernacular architecture was used, and the church was therefore a round house with a thatched roof.* In the early twelfth century the church was still standing, with an outer stone wall and within a large shed, when the respected historian William of Malmesbury (by 1125) saw this little "church of boughs" within Glastonbury Abbey, then a leading pilgrimage center for medieval Britain. William alluded to its pavement as made up of "triangles and squares" as follows:

> This church, then, is certainly the oldest I know in England, and from this circumstance deserves its name [Vetusta Ecclesia]. . . . In the pavement may be seen on every side stones designedly inlaid in triangles and squares, and figured with lead, under which, if I believe some sacred enigma to be contained, I do no injustice to religion.[2]

William would have entered the rectangular wooden hall that enclosed and protected the original church, but in 1184, the hall and little church burned down, leaving only the pavement. A rectangular stone chapel was then erected within two to three years, in the Gothic style, called the Lady Chapel (fig. 7.2, *bottom*); this chapel probably incorporated a vital dimension relating to the Old Church, and its rectangular structure was symbolic of the Virgin. From this, the overall size of the original church and its sacred geometry were deduced, largely by John Michell.

Hexagonal and Vesica Geometries

St. Mary's Chapel (the Lady Chapel) continued the Old Church's purported dedication to the Virgin Mary from early times, though thousands of church buildings were dedicated to her in the medieval period due to the then cult of Mary. This building was made symbolic of the Virgin Mary through the ratio of its longer side to its shorter side, a ratio equal to the square root of 3 (just greater than 1.732). Such a rectangle is significant to sacred geometers

---

*The structure, visualized by Frederick Bligh Bond in figure 7.2 (*top*), resembles the Chinese Hall of Light, whose shape was derived from a tower to observe the lights in the heavens (see Soothill, *Hall of Light,* chapter 10).

because, in the Gothic period, it was the basis for the ad triangulum design scheme. The Old Church was probably offset to the east (fig. 7.3, *right*) so that the path connecting the north and south doors (the path used by processing pilgrims) avoided crossing its sacred pavement. Like ad quadratum, this geometry permits the easy doubling and halving found in so many sacred buildings and could have been used to locate other buildings, such an outer ring (or diamond) of hermit cells, within successive doublings from the center.

The in-circle's diameter is half the out-circle's diameter, while the inside width of the chapel forms another circle, half again in diameter to the circle inside the hexagon (fig. 7.3, *left*). The square root of 3 is associated with the Virgin Mary by its ability to form a vesica piscis, using those arcs from the center of either side that have radii equal to the width of the rectangle

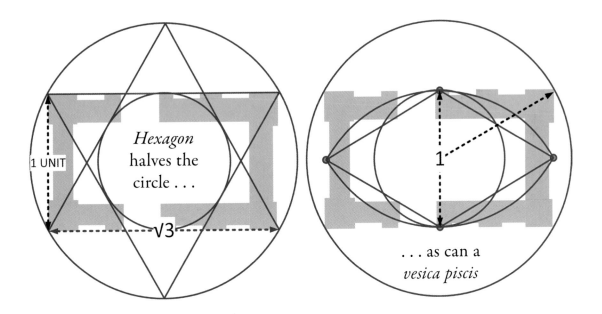

FIGURE 7.3. *Left,* hexagonal geometry between the out-circle of the rectangle and its width; *right,* an eye-shaped vesica is made from two circles, each having their centers on the opposite outer wall of the chapel's width. At Glastonbury, a common radius length of 39.6 feet defined the width of the chapel, and the two circles cross at the outside centers of the end walls, defining its length of just under 68.6 feet.

(fig. 7.3, *right*).* Jesus was called the Fish (*piscis*) because the ancient Greek for "Jesus Christ, Son of God, (Our) Savior" has initial letters spelling the Greek word for *fish*.

The vesica piscis shape is also familiar in Christian iconography as an aureole ("aura") or mandorla ("almond nut") that is used to encapsulate the images of either Jesus or the Virgin Mary. Furthermore, astrological significance was given to Jesus for starting the 2160-year-long precessional Age of Pisces, while Virgo, his mother, is the opposite zodiacal sign. The aura and mandorla were mysteriously connected through his birth. In *How to Read a Church,* Richard Taylor explains this as "Almonds are also associated with the Virgin Mary, because of their symbolism of divine flavor, the pure white of their blossom, and the womb-like shape of the almond's nut."[3][†]

The Lady Chapel expressed the specifically medieval Mary cult, but, most fortunately, its builders made the width of the Lady Chapel the same as an

Jesus in Majesty with Four Gospel Animals

Early Christian sign for Jesus

Our Lady of the Assumption
Taddeo Gaddi, 1350

FIGURE 7.4. Jesus (*left*) and the Virgin Mary (*right*), each inside a vesica, also known as a mandorla. Between them is the fish sign of the early Christians, with the Greek word *ichthys* within, meaning "fish."

---

*These geometries are both based on the equilateral triangle that was used for land surveying and cartography.

†The almond shape also represents the vulva, a yonic symbol (where *yoni* is the female equivalent for *phallus*). The vesica piscis (or "fish bladder") represents the birth passage of Mary and the divine feminine. It is associated with the waxing and the waning Moon phases, the Moon being a strong feminine element. Its link to the almond is its white blossom in the purity and holiness of this color. From WordPress, "Mythology and the Seasons," May 27, 2013.

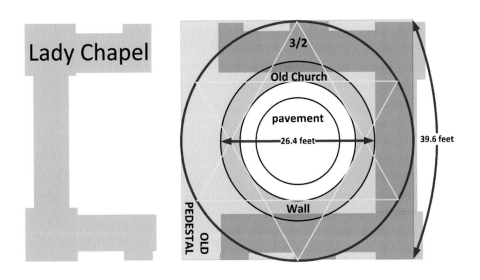

FIGURE 7.5. The Old Church is likely to have had a level pedestal that
was 3/2 of the church's diameter, a distance that was carried forward
in the width of the Lady Chapel.

earlier structure that protected the round church, the width of the Old Church's
square pedestal,* and from this generated the root-3 geometry relative to the old
rectangular building it replaced (fig. 7.5), according to St. Mary's vesica symbolism. This measurement of 39.6 feet for its width is a number native to the Equal
Perimeter Model, representing the mean radius of the planet as 3960 miles (see
chapter 3).

Anatomy of a Spiritual Center

In the interpretation I use here for the original Christian foundation at
Glastonbury I drew on historical records, similar sites, automatic writing (via
English architect and psychic researcher Frederick Bligh Bond), a Flemish
oil painting (via video producer Arthur L. Howe), and John Michell's books.
Michell conserved the details he found and developed them in his many books
on ancient mysteries spanning many decades, integrating the unique numeracy

---

*John Michell observed the similar situation of the circular church in the traditional Chinese Ming
T'ang, or Hall of Light, from Sinologist William Edward Soothill's book *The Hall of Light,* in
which he described the circular structure within a pedestal 3/2 times larger in side length than its
diameter, as per figure 7.5, *right,* giving in this case 39.6 feet to 26.4 feet for the circular church.

and significance of equal perimeter geometry—then sometimes called by him the foundation pattern but most often the New Jerusalem diagram.[4] Michell gives his sources for the New Jerusalem diagram as:

1. The precisely measured and published dimensions of the ground plan of Stonehenge, built during the third millennium BC.
2. The dimensions given in Plato's ideal city in *The Republic,* written about 500 BC.
3. Saint John's holy city, New Jerusalem, whose dimensions are given in Revelation 21 [written around 95 CE].
4. The dimensions, actual and legendary, of the Celtic Christian foundation at Glastonbury, the "English Jerusalem."
5. The mysterious revelation to F. Bligh Bond of the measurements in the Glastonbury sacred geometry.[*5]

Others adopted Michell's discoveries regarding Glastonbury's geometry and metrology to create their own interpretations. One of the most mysterious and helpful here was that of Arthur L. Howe,[†] whose 2007 film *The Painting and the Floor: The Glastonbury Code* borrows the outer measures, found by Michell, for the Old Church and its environs. By his own account, Howe was led to a book in his library on the Flemish masters, and an early-sixteenth-century painting called *The Annunciation,* by Bernard van Orley, as being a geometrically-encoded design for the original Glastonbury Pavement.

> In the following programme, the author explains his research into the historical origins of the foundation of Glastonbury Abbey, paying particular attention to the decoding of an anonymous Flemish painting *The Annunciation of the Virgin,* a painting that appears to have been purposefully crafted to encode the design of the sacred floor and the concentric details of the original church, reputedly built at Glastonbury by Joseph of Arimathea.[6]

---

*One notes that John Michell failed to indicate, to his *Glastonbury* audience, how seminal the Great Pyramid was in revealing the Equal Perimeter Model to him, as one sees in Michell's other books and Piazzi Smyth's *The Great Pyramid*'s delineation of that and the Equal Area Model (fig. 4.2).
†Howe passed away in 2016.

The destruction of Glastonbury Abbey, during the dissolution of the monasteries (1539 CE) by Henry VIII, included the ruination of the Lady Chapel and the loss of its pavement. Is it possible that a Dutch painting could have saved its essential measurement?

In figure 7.6, Gabriel and Mary both stare at an octagonal floor tile that represents Mary's womb, an octagon 24 units across, this being Mary's number as 3 × 8, co-adjacent with 26 (the letter-code value of YHWH when summed) to 25, that is, 5 squared. The number 5 (Plato's human creative number) when squared (see chapter 1) forms a Lambda-like diagram of the first five odd

FIGURE 7.6. *Left, The Annunciation* depicts Mary in her role as Virgin Mother; *right,* the diagram details Mary's symbols organized as the Pentacle of Mary within the painting (in units of half-remens), noting Mary's face as a vesica: *bottom left vertex,* lily in vase = purity; *right,* shears (Agnus Dei) and loincloth; *left,* Gabriel and wand; *right,* Mary's bed chamber; *top,* the divine world.

© *The Fitzwilliam Museum, Cambridge, used with permission.*

numbers: {1 3 5 7 9} which sums to 25 (1 + 9 = 10; 3 + 7 = 10; 10 + 10 + 5 = 25). In a right triangle, sides of 24 and 25 generate a third side of 7, the Royal number and hence Jesus, King of the Jews. As a ratio, 25/24 is the Just semitone—the signature interval of Just intonation found in the harmonic model leading to the twelve tones of chromatism, namely the twelve disciples or tribes of Israel.

As we shall see, further specific knowledge of the Glastonbury pavement was somehow preserved until Bernard van Orley (c. 1492–1542) composed *The Annunciation,* placing the encoded information into the quasi-public domain of art collections. It is well known that esoteric ideas and geometrical design principles came to underlie some post-Renaissance paintings. Secret societies and public interest in the arcane or occult knowledge was growing beyond ecclesiastical or secular control.

It is interesting to see how a painting could contain essential details enabling reconstitution of some ancient knowledge, just as Plato's dialogues have recently enabled diagrams of his ideal cities or of the island of Atlantis, not included in manuscripts, to be reconstituted for modern readers.

The Painting and the Floor

From Howe's film, one can glean the set of key measurements for this old round church that van Orley might have known. Howe silently employs some of Michell's work, which had already abstracted and deduced many important details, for instance, of the rectangular enclosure built for the Old Church* and of the surviving pavement's new Lady Chapel, which was built with a width of 39.6 feet, numerically equal to the mean Earth radius of 3960 miles.[7] This latter length is 6! (that is, 720) times 11, or 7920, twice 3960. It is one of many geodetic meanings found in the Equal Perimeter Model (see chapters 2 and 3). However, Michell had no concrete clues to sizes within the Old Church and the pavement's diameter, while Howe, via the painting, provides what these might have been. The Old Church's early-first-millennium design was probably the earliest Christian foundation in England.

---

*The diameter of the Old Church has been established as probably 26.4 feet. Michell (1990), 130–31.

TABLE 7.1

DIAMETERS OF THE OLD CHURCH

In English and Saxon feet, numerically unified as integers
by a third measure, the half-remen of 0.6 feet

Diameter	English feet	Saxon feet	Half-remen	Numerically
Outer Limit – 11	79.2	72	132	*mean Earth diameter*
Outer Chapel – 14	50.4	—	84	
Outer Chapel – 11	39.6	36	66	*mean Earth radius*
Roofline of Church	28	(80)		*(88-feet circumference)*
Old Church	26.4	24	44	*25 latitudinal feet of 1.056 feet*
Inner Diameter – 14	25.2	—	42	
Inner Chapel	22.8	—	38	
Inner Church – 11	19.8	18	33	
Wattle Wall	18	—	30	
Pavement	14.4	—	24	
Cross	7.2	—	12	
Moon – 3	5.4	—	9	*3 half-remen units are Assyrian double foot*

Table 7.1 shows the actual diameters of concentric circular features proposed by combining the data of both Michell and Howe. Two measurement units were mentioned by them, the English foot in fractional lengths such as 39.6 feet (198/5) and the Saxon foot of 1.1 feet (11/10), then in integer lengths.

A. The English foot lengths provide the canonical numbers within the Equal Perimeter model, these being associated with the size of the Earth and of the Moon, in miles, numerically scaled down by 100 and then in feet rather than miles. The scale was therefore 1:528,000.

B. Only some of the diameters can divide by Saxon feet since those diameters are ideally 11 units in the model. Thus, 3960 becomes 39.6 feet, which is equally then 36 Saxon feet.

Saxon feet will not divide into the diameter 50.4 feet of the circle of equal perimeter to the square of 39.6 feet. It is the prime number 11 in 11/10 that cannot be cancelled within 50.4 feet. But there is a unit of length that can divide into all of the diameters to render them integers: the half-remen of 0.6 feet (3/5):

- 39.6 feet is 66 half-remens, and
- 50.4 feet is 84 half-remens

The half-remen of 0.6 feet (3/5) places the design of the Lady Chapel, the Old Church, and its pavement into a single continuum of whole numbers, as in Table 7.1 on the previous page.

Using feet, the numerosity of the mean Earth and the Moon's diameter were preserved, while the half-remen enabled the whole scheme to be stated as integers, even at this small scale.

This may explain why Michell found Stonehenge (fig. 7.7) was built on the same model and at the same scale as the Old Church (fig. 7.8). While the Druids

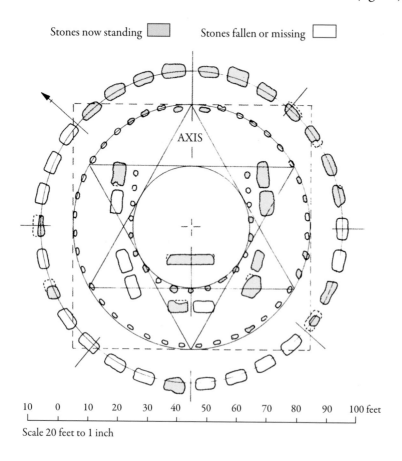

Stones now standing ▢   Stones fallen or missing ▢

AXIS

FIGURE 7.7. The Lady Chapel within Stonehenge.

From John Michell, The Dimensions of Paradise, *32.*

10 0 10 20 30 40 50 60 70 80 90 100 feet

Scale 20 feet to 1 inch

may have remembered the model of Stonehenge, the early Christians might have known the same model from another source. Both monuments may have been the same size because the model was always best stored (in an integer-hugging tradition) as half-remen, to remember all the diameters. This scale produces as small a monument as is possible while still maintaining an integer description for it. New concentric features of the Old Church could then be made as integers using the half-remen as a building unit. English feet would then reveal the cosmological numbers behind the integers, and it was probably for this reason Michell found the Lady Chapel plan corresponded to the plan of Stonehenge based on the length of 39.6 feet and the Lady Chapel's implied vesica.

Table 7.1 can now be projected onto the Lady Chapel with north at the top (fig. 7.8).

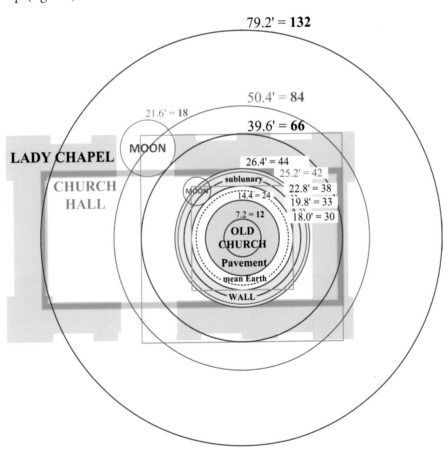

FIGURE 7.8. The Lady Chapel with the round Old Church and intermediate Old Church Hall, which shared the inner width of the chapel.

The model relates to both the cosmic reality of the macrocosm (the sublunary world) in circular form and, within this, a model of the microcosm that was placed as a pavement. This was why Michell said the outer circle of the sublunary world (14 in diameter; here, 50.4 feet) should express the numerosity of 3168 (44 × 72 when π = 22/7) in their scaling to define the Moon as the perimeter of a holy space containing models of the mean Earth (11 in diameter; here, the 39.6 foot width of the Lady Chapel). The Greek word *temenos,* for "a sacred space," could be any defined boundary, but Greek and other traditions moved toward built structures such as temples that were square or rectangular like the Heraion, sacred groves, synagogues, churches, mosques, and so on. These started, like the wattle church, as vernacular houses but were soon enlarged and extended, remaining connected to these perimeters through the use of a numerical canon and related metrology.

The idea of the microcosm collides with the people within sacred buildings, one being an archetypal class and the other an instance of its expression. At the very least, worship of the sacred within special spaces was foreordained in early religious structures, connecting them to heaven through conforming to a cosmic reality.

The Glastonbury Code

In the *Annunciation* painting, the archangel Gabriel is appearing to Mary, transmitting the plan of Life for the advent of Christ to her womb. Figure 7.9 shows the pattern seen by Howe, with Michell's foundation model having a radius for the Moon of 9 half-remens, a sublunary diameter of 42 half-remens, and the mean Earth of 33 half-remens. It is as if Christianity is portrayed as being founded according to a pattern.

1. Gabriel is in the sublunary world (see also point 6) with his hand held up, fingers (in Pantocrator gesture*) defining a horizontal that strikes the distant window.
2. A vertical, down the axis of the far window, travels to the central irregular octagon at bottom to define the vertical axis of the foundation.
3. The horizontal axis, defined by (a) a ring on Gabriel's wand, (b) the bases

---

*As seen with Jesus when shown as Lord of the World and in Byzantine deisis panels.

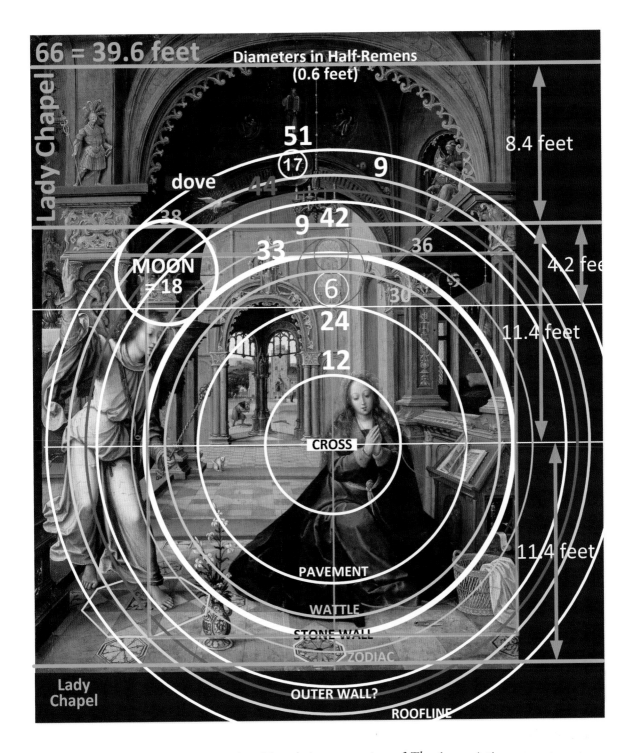

FIGURE 7.9. Arthur Howe's interpretation of *The Annunciation.*

Image underlay © The Fitzwilliam Museum, Cambridge, used with permission.

of the distant arch's pillars, then running through (c) Mary's prayerful hands and (d) the top of the reading stand, is the axis of the Lady Chapel.

4. Gabriel's left hand on the wand defines the wattle wall of the Old Church (30 half-remens), while the end of the wand touches the edge of the pavement (24 half-remens), the wand then tangential to the pavement.

5. The inner circle (12 half-remens) contains Mary's face and gaze, in the style of mother and child, yet instead of Jesus, the gaze is to the womb-like irregular octagon pattern of the tile at her feet. Gabriel's gaze is likewise focused on that same octagon tile, expressing that both have a common purpose.

6. Gabriel's head, torso, right leg, foot, and the focal octagon are all delineated by the stone wall of the circular church (as the Moon), the inner wall (33 half-remens), and the outer wall (42 half-remens), with circles also containing signs of Mary's purity (the vase and lilies) and Christ's death (sheep shears and discarded garment). The symbolic theme is Mary's purity and Jesus's birth and death as a sacrificial lamb.

7. The outermost circle (51 half-remens, or 3 × 17) contains a scene with a diameter that touches the far edge of the Moon circle (9 half-remens), which has the basic value 17 in this geometry, a number shown clearly by the dish hung on the wall like a clock, having seventeen radiants (fig. 7.10).

FIGURE 7.10. Blown up portion at the top of the design's (51 units) proposed outer circle of the perimeter model. The painted brass dish has 17 radiants, the correct number of units within the Equal Perimeter Model, since 3 × 17 = 51.

© *The Fitzwilliam Museum, Cambridge, used with permission.*

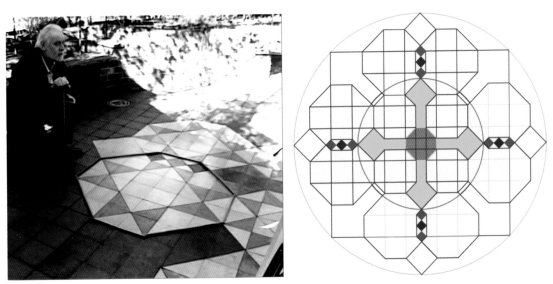

FIGURE 7.11. *Left,* the late Arthur L. Howe and his reconstruction
of the Glastonbury Pavement at a friend's garden.

Photo by Nick Curtis, used with permission.

Right, my own drawing of his design, based on an irregular octagon
in the center expanding twice through the square root of 2,
in four cardinal directions. The diameter is 14.4 feet, or 24 half-remens.

8. A dove can be seen, at top left in the painting, on what Howe surmises is the roofline diameter, signifying the end of the Flood or, in reverse, the roof of the church as ark, a circle 88 feet in circumference, or 80 Saxon feet.

9. The Lady Chapel had walls 8.4 feet thick (14 half-cubits), and there are 22.8 feet (38 half-cubits) inside, between the walls.

10. The Old Church's pavement, which the chapel was designed to wrap around, is accommodated with 14 (38 − 24) half-cubits to spare, allowing a 4.2-foot passage (7 half-cubits) on either side of the pavement.

The Human Reflecting the Cosmos

The centers of Christian pavements might have presented Christ as the microcosmos, exalted as the Cosmic Individuality, while being based on the much older tradition of the Equal Perimeter Model since the Earth and the Moon are the scenes of the human drama. Christianity inherited ancient models and traditions, including those of the Bible, dynastic Egyptian, Persian Zoroastrianism, and Near Eastern matriarchal forms, which were then reflected in art and metaphor. The high-status role of pavements within churches can be seen, in those that survive, as representing central ideas that would become the liturgical canon of Communion. An example is the spectacular Cosmati pavement at Westminster (see the next section, "Westminster's Cosmati Pavement"), which was a confluence of British, medieval, and Italianate traditions.[8] Unusually, there is an inscription containing a well-known type of cosmological rhyme about the age of the world, and the pavement's geometry powerfully expresses the Equal Perimeter Model. The unique form of the Cosmati pavement implies that the Italian Cosmati master (called Peter) may have incorporated some British design elements into his pavement, especially since the Glastonbury setting of the Old Church, as reconstructed previously, can easily be drawn as an Equal Perimeter Model, as in figure 7.12.

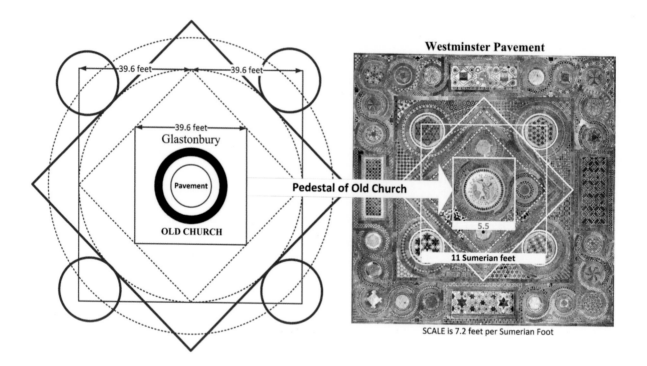

FIGURE 7.12. Comparison of the Glastonbury Old Church completed as the Equal Perimeter Model of the Earth and the Moon (*left*) and the Westminster Cosmati pavement, whose center mirrors the Old Church (*right*) at an exact scale of 7.2 feet per Sumerian foot.

© *Dean and Chapter of Westminster, used with permission.*

The scale between the two pavements shown in figure 7.12 is exactly 7.2 feet at Glastonbury to each Sumerian foot (of 12/11 feet) at Westminster (6.6 to 1), where the whole design was 24 English feet, this length *also* being 22 Sumerian feet (of 12/11 feet), allowing the mean Earth diameter in the Westminster model to be 11 Sumerian feet. At this scale, the border of the central Westminster roundel (fig. 7.13) signifies the size of the Old Church and, visually, its original wattle wall. The central gemstone disk of Westminster could then be the 14.4-foot diameter size of the original Glastonbury pavement, now 2 Sumerian feet across.

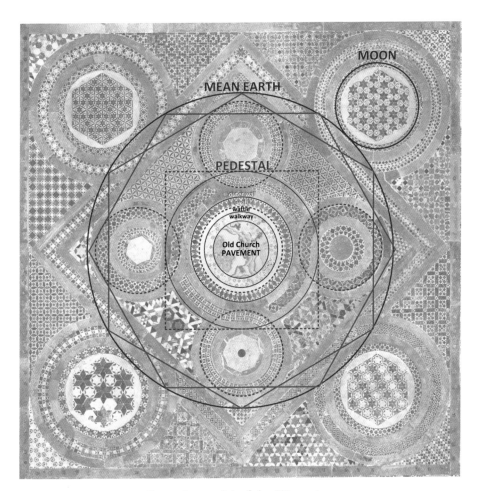

FIGURE 7.13. Central roundel of the Westminster pavement as Glastonbury Old Church, its own pavement and original square platform with four roundel "gates."

© *Dean and Chapter of Westminster, used with permission.*

FIGURE 7.14. Photo of the Cosmati pavement at
Westminster Abbey after its recent renovation. North is upward.

© *Dean and Chapter of Westminster, used with permission.*

WESTMINSTER'S COSMATI PAVEMENT

Westminster Abbey was built by Henry III and dedicated to the Saxon king
and Saint Edward the Confessor. The pavement there was primarily used for
the coronation of subsequent English, then British monarchs. Finished by
1268, the Cosmati craftsmen* used their signature style of mosaic roundels,

---

*The Cosmati craftsmen probably expressed many influences, including those of the Comacine
masters of Lake Como, the Byzantine mosaics, and the floors of the Roman empire.

which suited the Equal Perimeter Model. Most examples of Cosmati work are found in Italy and Rome, though none are of this design or complexity.

The diagram realized was ancient: the Equal Perimeter Model found throughout the then-known world, including Britain, at Stonehenge, but St. Mary's Chapel, built at Glastonbury just over eighty years before the Cosmati pavement, held the then surviving pavement of the Old Church that had burned down. It is possible that the Westminster pavement sought to reference the Glastonbury pavement since it was the oldest church in Britain.

The widespread distribution of this model in sacred art is hard to explain. Craft guilds such as the Cosmati might have been familiar with the pattern, but the model may have been known to the Celtic Church and hence in the Celtic sacred center of Glastonbury BCE. The Anglo-Saxon people had a strong "dark-age" tradition of art and diagrammatic design. This can be seen as finds within hoards of jeweled treasure, illustrated books such as that from Kells, the forms within Celtic knotwork, and the habit of annotating schemata as diagrammatic marginalia within hand-copied and otherwise text-only manuscripts.

The whole design was registered using three black, square keystones (fig. 7.15). These are shown between the edge of the pavement and the white frame for the design, within the added white circles. The side length of the outer square (shown with a white border) was then 24 feet (the conventional view must be in English feet), but this is also 22 Sumerian feet* of 12/11 feet, enabling simple and direct representation of the pattern in which the inner circle, representing the mean Earth, has a diameter half this, 11 Sumerian feet. Division by 11 and multiplication by 12 turns feet into Sumerian feet, this being the native unit for this model. The out-square would be 44 Sumerian feet (also 48 English feet). A 14-unit diameter then gives a circle of equal perimeter, that is, $14 \times 22/7 = 44$ Sumerian feet. The diameter of the Moon must then be 3 Sumerian feet. If the mean Earth *radius* is taken to be 3960 miles, which, divided by 11 is 360 miles, then multiplying that times 3 gives 1080 miles; C. W. Allen, in *Astrophysical Constants*, gives the mean *radius* of the Moon to be 1080.067 miles, a figure accurate to one part in a thousand!

---

*The Sumerian foot was famously employed in the Great Pyramid as its truncated height of 440 Sumerian feet, then also 480 English feet.

FIGURE 7.15. Superimposition of the New Jerusalem pattern
over the Westminster pavement, based on obvious similarities
of the roundels and central diamond setting.

© *Dean and Chapter of Westminster, used with permission.*

In figure 7.16, the circle of diameter 11 Sumerian feet fits within the diamond
as its *in-circle,* representing the mean circumference of the Earth within the pavement. It is perhaps no accident that Stonehenge, Avebury, and Westminster Abbey
all lie within the mean degree of latitude 51–52 degrees north, where the parallel
of latitude equals that of each of all the 360 degrees the mean Earth *would have*
(364,953.6 feet) if the Earth did not spin and were a perfect sphere.

The outermost circle (diameter 14 Sumerian feet) passes through the center of all the four roundels outside the diamond. The roundels represent the Moon, whose diameter is then 3 Sumerian feet to the mean Earth's 11 Sumerian feet. If the roundels were to roll around, just touching the inner circle, the path of their centers would therefore describe the outer circle of diameter 14.

Michell correlated this strange proximity of the Moon to the Earth as representing the sublunary sphere, considered in medieval times as defining the limits of the influence of the Moon within a geocentric model with the Earth as its center and within the model of equal perimeter. The central roundel might be the Earth shown in miniature, surrounded by sublunary stars in a band repeating the outer circle in miniature. As noted in the previous section

FIGURE 7.16. The application of repetition by scale within the pavement.

© Dean and Chapter of Westminster, used with permission.

on Glastonbury, the central roundel is 2 Sumerian feet in diameter and, at a scale of 7.2 feet to a Sumerian foot, the central porphyry disc could represent the size of the now lost Glastonbury pavement of the circular Old Church. If this is true, then the pattern of the lost Glastonbury foundation was chosen to be the subject for the Cosmati masters' work.

The Microcosm within the Pavement

Medieval Christianity was strongly influenced by Classical Greece and Pythagoreanism. Pythagoras pioneered a type of holistic perception about the structure of the world based on numbers, and this, alongside the work of the pre-Socratic philosophers, triggered the theory building of the Classical world called natural philosophy. The earliest Christian monasteries preserved parts of this thinking, but the Arabic thinkers of the late first millennium BCE, such as Al Kindi, had further developed it by the end of that millennium. Spain, Sicily, and the Holy Land gave Christianity contact with the new Arab and older Greek texts. The Greek model of Four Elements (earth, water, air, and fire) appears reflected within the pavement's Equal Perimeter Model (fig. 7.17). The human as microcosm lives in the sublunary world where the unique planetary world is made out of the Four Elements.

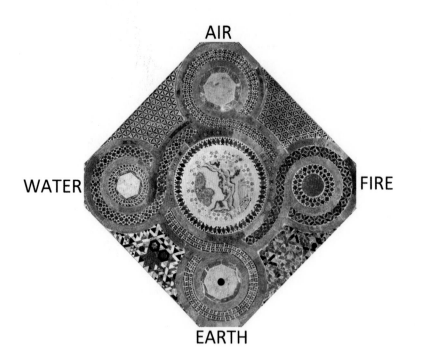

FIGURE 7.17. The sublunary world of the microcosm as the Four Elements.

© Dean and Chapter of Westminster, used with permission.

The Four Elements recognize the wonderful organization of solids, liquids, gases, and fire on the Earth, without which there could be no intelligible life or the immense beauty on the surface of the planet. These states of matter are perfectly scientific as the phases of matter but have largely lost their meaning today, but they were fully accepted and readable within the Westminster pavement by the educated classes of the thirteenth century and were simple enough for the common man to understand, being a palpable truism.

The Four Elements could today be construed as arising from our large Moon and its sublunary sphere, exactly because these states of matter have become so richly developed in their manifestations on the Earth, due to this dynamism. The Four Elements can also be seen within the harmonic relationship of the Moon to the outer planets. According to the harmonic cosmos, the traditional numbers associated with the elements can be seen within the Lambda diagram of Pythagoras, a simpler version of the astronomical matrix model of chapter 5. This astronomical matrix (fig. 7.18), first presented in my book *The Harmonic Origins of the World,* can reduce the harmonics between the outer planets and the lunar year to a single harmonic register of Pythagorean fifths, using lunar months to quantify the matrix. The 32-lunar-month period, realized at Le Manio and used at Crucuno circa 4000 BCE, is exactly 945 days long. One month is then 945/32 (29.53125) days long.* Ernest McClain's harmonic mountains (see fig. 7.18, p. 168) can be used to show 32 lunar months as 5/4 of 2 Saturnian synods and twice that, 64 months, equal 5 Saturnian synods.

Using the lowest limit of 18 lunar months, the commensurability of the lunar year (12) with Saturn (12.8) and Jupiter (13.5) would have been "cleared"† to integers using tenths of a month, revealing Plato's World Soul register of (6:8::9:12), but shifted just a fifth to (9:12::13.5:18), perhaps revealing why the Olmec and later Maya employed an 18-month "supplementary" calendar after some of their long counts.

By doubling the limit from 18 lunar months to three lunar years (36 lunar months), there is no need to clear the 13.5 since 2 Jupiter synods are 27 lunar months, so the lunar year can simply be doubled to 24 lunar months and the

---

*This is accurate to one part in 45,000 (less than 1 minute) when compared to the actual lunar month of 29.53059 days.

†Clearing multiplies a set of fractional numbers to eliminate their denominators and leaving only a set of integers.

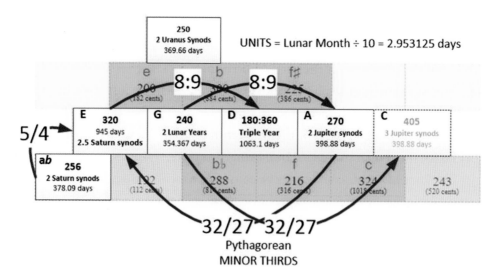

FIGURE 7.18. The elimination of 5 as a factor in the harmonic mountain for 36 lunar years, resolved using matrix units of one-tenth of a month and a limit of 360 units, each one-tenth of a lunar month.

32-lunar-month period is then within the register E-G-D-A of figure 7.18, while 5/2 Saturn synods (2.5) equal 32 lunar months and five synods, 64.

The Number Values of the Four Elements

By the medieval period, the numbers on the base of the Lambda diagram (fig. 7.19, *left*) were used to signify the four elements of fire, water, air, and earth, and these elements were shown separated, one from the next, by musical fifths—as Plato describes in the *Timaeus* creation story. However, in the decorated crypt of the Pope's summer palace, the Anagni Cathedral* (fig. 7.19, *right*), this allocation of numbers to the Elements differs but is, I believe, correct, with earth equal to 8 and fire equal to 27.

In the planetary matrix of synodic periods, Saturn is distinguished as the cornerstone, having no other factors than the powers of two with respect to the lunar year. Its 12.8 month synod is rationally commensurate (5/2) with the 32-lunar-month period, which is 8 merely doubled twice to fit within the

---

*"The numbers assigned to the Elements are the reverse of the ordinary sequence, Fire being given the number 27 and Earth, 8." Foster, *Patterns of Thought,* 135. However, this unique "mistake" restores the sequence of musical fifths.

octave limit of 36. It therefore seems that Saturn represents the earth element. It is quite traditional for the Moon and lunar year of 12 lunar months to represent water and half of the triple lunar year, 18 months, the element of air. The element of fire is then the double synod of Jupiter (27), again in lunar months.

8: Since 32 lunar months is 945 days, then the Earth that continually rotates toward the east is Plato's "cycle of barrenness," which is associated with the prime number 2 of octave doubling, an octave being seen as a container awaiting intervals, as with a womb that enables children to arise, but only if fertilized by a male (or odd) number. Saturn was called slow, weighty, and ponderous, and this represents the solid state of objects as the traditional giver of definite boundaries, appropriate given that Saturn is also the outermost visible planet.

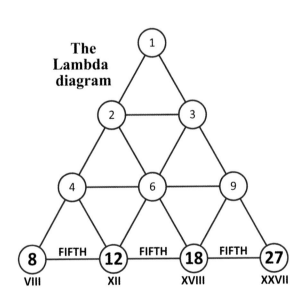

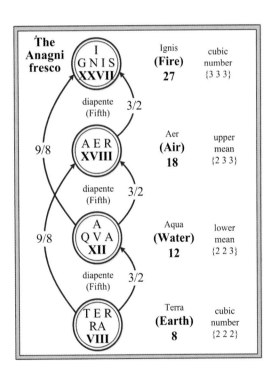

FIGURE 7.19. In the crypt of Anagni Cathedral is a schematic in which customary allocation of numbers to Elements is reversed, namely, 27 and 8 (*right*)—the correct order is shown on the base of a Lambda diagram (*left*). The natural order of the Lambda diagram restores the relationship of successive fifths between the Elements.

12: The Moon's links to water come directly through the tides as the first of many traditions, such as that it fills with water until "full" and thereafter starts to "empty"—in tidal fashion.

18: The lunar year also collaborates with the near-anniversary of 3 solar years, equaling 37.1 lunar months. This periodicity was studied by the megalithic astronomers of Carnac as a right-angled triangle whose counted base was 36 lunar months: there were 36 stones in this kerb at the Le Manio Quadrilateral.

27: After two further loops of Jupiter, the middle of Jupiter's retrograde loop will be punctuated by a full Moon (at maximum retrograde), the Sun then in line with the Earth and Jupiter (see "The Harmonic Planetary Model," p. 102). This fact enabled the loops of outer planets to be counted in lunar months, as required. The association of Jupiter with fire seems natural since his primary weapon is lightning, the cause of natural fires.

The Mandala Effect

The Westminster pavement resembles the geometrical art found in a number of religions. When its geometry is laid over a Buddhist mandala, the interplay of squares and circles defines the composition surrounding a central image, as in figure 7.20.

Mandalas seem to express spiritual centers using the same sacred geometries as pavements and domes. The practical geometry of quadrature enables the diamond within a square to make a square that is half as big as the original. In-circles and circles equal in perimeter to that square form annular rings on which visual objects (such as roundels) can be placed. These are then like moons relative to the central image.

A mandala, like a temple, has the microcosm at the center. In the Westminster pavement, the Equal Perimeter Model populates the sublunary region with the Four Elements and gives a central repetition of the globe, as a central disk, that may hark back to Britain's earliest Christian foundation, the round wattle church of Glastonbury. As such, the homespun secrets of Glastonbury were appropriately placed within the Cosmati coronation pavement for the kings and queens of what would become a worldwide empire.

STATIONS
OF THE
MOON

EQUAL PERIMETER MODEL

EQUAL AREA MODEL

THE MEAN EARTH

THE MICROCOSM

THE SUBLUNARY WORLD

FIGURE 7.20. A mandala conforming to the same design elements as the Westminster pavement. In the center of this painted seventeenth-century Tibetan "five deity mandala," is Rakta Yamari (the Red Enemy of Death) embracing his consort Vajra Vetali; in the corners are the Red, Green, White, and Yellow Yamari.

Courtesy of Rubin Museum of Art.
Note the wide black circumference of the Equal Area Model of the
nodal period relative to 33 years. (Added diagramming is in white.)

CANTERBURY'S COSMIC CONUNDRUM

This mosaic pavement of the Canterbury Cathedral (fig. 7.21) was given to the cathedral by the pope of Rome, who had works and materials from many ancient buildings in storage. These were stored in Ostia, near Rome, "until required for a suitably deserving project." According to Colin Joseph Dudley, a British geometer, "The Canterbury pavement, like the pink marble used in the arcade surrounding the Becket shrine would have enjoyed such a history."[9] But then

FIGURE 7.21. Mosaic pavement before the shrine of Thomas A. Becket, in the Cathedral Church at Canterbury.

Drawn, restored, and engraved by William Fowler, Winterton, Lincolnshire, pub. 1807, British Library, London, UK, copyright British Library Board. All rights reserved/Bridgeman Images.

how old is it? Dudley provided a likely answer based on the ancient availability of the rare gemlike stones of red and green porphyry, their rounded, not flush, setting in the pavement, and the metal framework on which the gemstones were set, like jewelry. The Westminster Cosmati pavement of a century later, like all medieval pavements, has flush decorative stones, few or none of them being porphyry because, says Dudley, new green porphyry was unavailable from its only source in Egypt well before 79 BCE and red porphyry from before the time of Emperor Justinian, who built the Hagia Sophia in Constantinople. As these stones became rarer, they had to have been sourced from earlier buildings, or other stones were used instead, such as olive-green serpentine or red marble.

This large pavement may therefore have come from the early centuries BCE, when both green and red porphyries were still available. The 3-millimeter brass strip frame on which it was built made it portable—possibly when being moved from its original setting but also as a technique for its accurate construction in the first place.

The Models Within

The pavement's unique geometrical form, of four large roundels symmetrically punctuating each side of a square diamond, is a subtle form of the Equal Area Model, between the square of 12.8 feet (see fig. 7.22, p. 174) linking the extremities of the roundels and the circle that would touch the eight junctions of the four roundels with the diamond, also of 12.8 feet.

The Canterbury pavement therefore relates the square, divided by the 33 years of the solar hero, to the circle of radius 18.618 years of the nodal cycle as its outer context. But this relationship is now shown to be somehow dominated or defined by the Saturn synod of 378 days, which, times 18, equals 6800 (18.618 years) of the Moon's nodal period + 4 days.

To achieve the *radius* of the nodal circle, one must divide the square's side length (12.8 feet) by 33 and multiply that unit by 18.618. The width of the square being 12.8 feet reveals the unit as being 0.3̲8̲7̲ feet, which times 18.617 (in years, the actual 6800 days of the nodal cycle) gives 7.22144 feet as the radius from the pavement's center, which is required to punctuate the diamond.

The centers of the roundels in the pavement are defined by the corners of the 8-foot-side-length square, which defines the inner dimension of the microcosm and also the inner border of the diamond. These corners are set as 4 feet

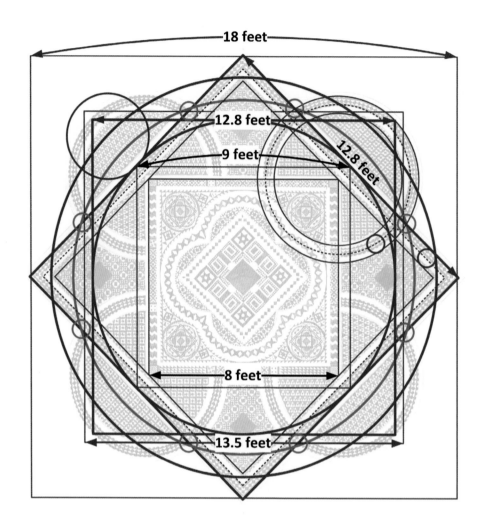

FIGURE 7.22. The metrology within the Canterbury pavement, which conforms to the Equal Perimeter Model but with innovations integrating the Equal Area Model (*red*) and the Harmonic Model.

*Image underlay from the British Library, London, UK, copyright British Library Board.
All rights reserved/Bridgeman Images.*

times the square root of 2 from the center, allowing the basic form of the outer part of the pavement to be drawn by the original craftsmen. As is normal in the Equal Area Model, irrationals are subsumed in their practical solution, in this case (a) the root 2 implicit in the roundel centers and (b) the reciprocal of *phi* (1/phi = 0.618) within the 18.618 radius.

Drawing radiants from the center to the eight junctions of the nodal period with the outer diamond, one sees an irregular octagon emerge that can be rendered as one of those cross shapes similar to the Maltese cross (fig. 7.23). Crosses are not unique to current-era Christianity; they emerged naturally out of fourfold structures, including the fourfold structures of time, the cardinal directions, and specific models like that of equal perimeter, the Earth, and the Moon.

The Equal Area Model of the outer pavement is combined with an *exact version* of the harmonic model in lunar months, another time-based model between the lunar year and synods of the outer planets, these forming the border widths.

The side length of the pavement's square extent is 18 feet, while the side length of the diamond, ad quadratum to it, is 45/32 of that, or 12.8 (64/5) feet—the number of lunar months in a Saturn synod, while $\sqrt{2}$ would be 12.73 feet. This highlights the cubit of the lunar year of 18 lunar months, and the half-cubit of 9 months, as an octave within which the outer planets form inner notes. In this case, Saturn is the tritone and as such is dominant as a diamond between the 18- and 9-wide squares.

FIGURE 7.23. The junctions between the outer diamond and the four roundels sketch a Maltese cross from the center of the Canterbury pavement.

Image underlay from the British Library, London, UK, copyright British Library Board.

As shown in figure 7.24, if the tonic of both of these is the note class D, Saturn is opposite D in the tone circle as A♭: in the clockface-like tone circle, Saturn = A♭, is at 6 o'clock relative to the tonic at 12 o'clock—just a semitone beyond the lunar year of 12 lunar months. It is a remarkable attribute, implicit in ad quadratum, that the harmonic consequences of √2 imply that the complementary squares belong as octave notes larger and less than it when turning the harmonic model into geometrical form. And here this appears to go beyond Saturn while presenting Saturn as the limit of the visible planets and hence defining the boundary of the planetary system as its diamond tips that touch the 18-lunar-month square (fig. 7.25). To show Jupiter, a longer synodic period (398.88 days)

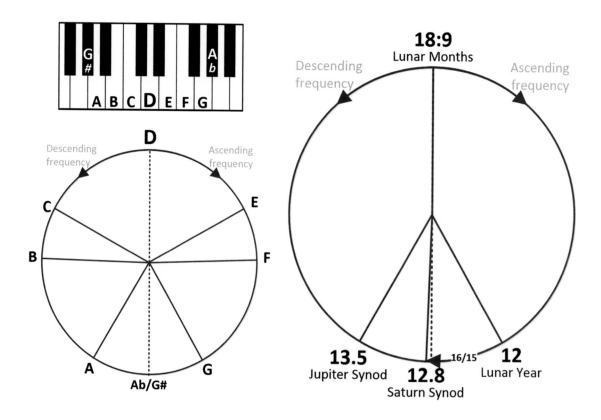

FIGURE 7.24. *Top,* the familiar keyboard of notes is symmetrical about both D and A♭/G♯; *left,* an axis of symmetry within the logarithmic tone circle for any octave; *right,* the same pattern is formed by the outer planets Jupiter (fifth) and Saturn (45/32 as A♭)—the tritone to D, which is 9 to 18 lunar months.

that is a whole tone beyond the lunar year (12 months), a square was used with a side length 13.5 lunar months, that is, 13.5 feet. This will exceed both the roundels and diamond/square that are the limits of Saturn, as is shown in figure 7.25 and by the circle around the pavement and by the cross in figure 7.23.

Jupiter's number (4) and sign (♃) derive from its four visible satellites, Io, Europa, Ganymede, and Calypso, and in the pavement, the roundels counterpoint the tips of Saturn's diamonds, symbolizing Jupiter with the four roundels (fig. 7.25). The eight important junctions between the roundels and the diamond are again visited, not by the nodal year circle but by the square of Jupiter's synod as each side's length (in purple).

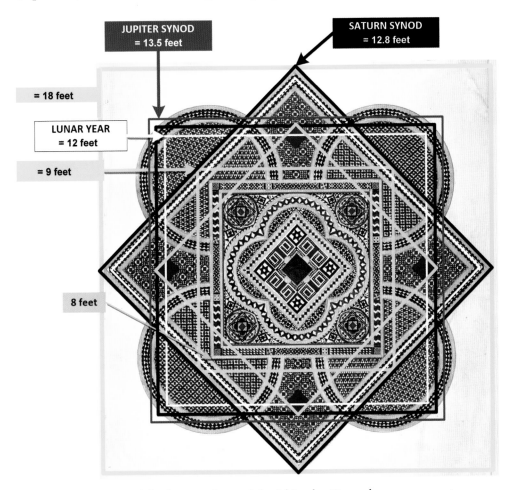

FIGURE 7.25. The harmonic model within the Canterbury pavement.

*Image underlay from the British Library, London, UK, copyright British Library Board.
All rights reserved/Bridgeman Images.*

The outer border of the pavement is a diamond within a square setting, in which 16 squares can be imposed (fig. 7.26). Traditionally, the human world called the microcosm is shown *within* the outer border of these models of equal perimeter, as at the Westminster pavement. The central four squares are 9 feet across, while the microcosm appears in an 8-foot square, and this is a major theme within the pavement of Jupiter's synodic ratio to the lunar year of 9 to 8.

Whether the pavement is BCE or CE, this is still after Pythagoras, who initiated the Western tradition that discriminated the cosmos and the bodily experience as a complementary and harmonious unity and developed the words *microcosm* and *macrocosm*. We experience the cosmos through our skies and our senses. So, a simple way to define the microcosmos is to picture it as held within the all-around horizon view of the heavens, exactly as the mega-

ring of 68 diamonds

range of the Moon's extremes about the solstice sun

FIGURE 7.26. The microcosm within the pavement; *left,* the whole design as a yellow square; *right,* enlarged as the structure being given to the microcosm within the macrocosm.

Image underlay from the British Library, London, UK, copyright British Library Board. All rights reserved/Bridgeman Images.

lithic astronomers had done. Figure 7.26 shows this as an area with a new type of content, unlike the simple repetition there of roundels and circles. To see how the pattern was geometrically meaningful, there must be something recognizable.

This portrait of the microcosm has the self as a central black diamond with four light-gray semicircles adjacent to each side looking at 45 degrees to the horizon, each looking toward a square in which the extreme solstice Sun is shown as a red rectangle facing the central black diamond. The square surrounding this Sun at the solstice has a range similar to the extremes of the Moon over its nodal cycle of 18.618 years. All this is stylized to suit the quadrature angle of 45 degrees, which is only actually the case at the latitude of northern England.

We have seen that the four circles punctuating the sides of the main diamond of the pavement are emblematic of the nodal cycle, which causes the cycles of

FIGURE 7.27. *Right,* the central region of the Canterbury pavement with the quadrature established for equal area only after (*left*) taking each quadrant to be the horizon viewed toward, for example, the northwest extreme events of the Sun and the Moon.

eclipses, the most powerful of which is the Saros cycle of just over 18 years. The only low-tech way to follow the nodal cycle is to count days since the cycle is accurately 6800 days long. There is a diamond border of 68 diamonds around the observer, implying that a count of 68 hundred-day periods will track the extremity of the Moon in the lunar orbit relative to the solstice points on the horizon, of which there are four. Beyond that diamond lies a double border with sixty crossed stars having four "leaves" of fifteen each, perhaps the measurement of the azimuth in regions of 6 degrees each, there being two stars within the range of the Moon on the horizon, that is, 12 degrees, during the nodal cycle.

The squares around the extreme Sun are worth quantifying also. Each side has a petal with eleven flames adding to 44, the size of the outer circle of equal perimeter, presumably to that square. The square border, itself, is black with forty-eight dots in its perimeter, while within it are thirty-six flames in four more petals. In a symbolic mode of quadrature, a further black square encompasses the maroon Sun square (on the horizon) and is surrounded by twelve black dots. I suggest the 12 lunar months of the lunar year are leading to **36** months (3 lunar years) and **48** months (4 lunar years) in an implied frame of **54** lunar months as a harmonic limit that is 4 synods of Jupiter. To finish this requires the observation that in 32 lunar months there are 945 full days, the magic formula discovered at Carnac, within which what became the Four Elements was found, in which each element was a musical fifth from those adjacent, as shown in Table 7.2.

TABLE 7.2
THE LIKELY FORMULATION OF THE
FOUR ELEMENTS IN LUNAR MONTHS

Calculations	Lunar Months	Element
starting from	32	Earth
$32 \times 3 = 96/2 =$	48	Water
$24 \times 3 = 72/2 =$	36	Air
$36 \times 3 = 108/2 =$	54	Fire

Therefore, the later Cosmati pavement at Westminster (as noted in the previous section) also showed the Four Elements within the microcosmic area.

8

The Focal Buildings of Islam

THROUGH ITS FIVE DAILY PRAYERS, Islam achieved a larger community of prayer through many troubled centuries, and these prayers were focused toward Mecca, the spiritual center of Islam, the Prophet's hometown, and toward its Kaaba in particular.

THE KAABA

The Kaaba is a plain building wrapped in a gold-and-black cloth covering called the *kiswa*. Though its shape is called a cube, it is actually rectangular in its proportions: 13 long, 11 wide, and 15 high. It was rebuilt recently to replace an older edifice, but its shape was probably developed when the Prophet was in his thirties. Five to ten years later, his religious mission was to create Islam at the behest of the archangel Gabriel, who had also enunciated Jesus's conception in Mary's womb many centuries before.

FIGURE 8.1. The Kaaba in 2019.

Photo by Muhammad Mahdi Karim for Wikimedia Commons.

The archangel Gabriel initiated a stream of revelations that, written down, formed the text of the Qur'an and the Hadith. Islam was to be a new dispensation, but, like the Kaaba, it was built on older foundations: the line of Abraham, leading patriarch of the early Jewish Bible of five books: the Pentateuch. In this way, Islam inherited many of the Bible's earliest narratives while, in principle, accepting those of the Jewish and Christian faiths as people of the Book, though different.* The early Bible was written by literate Jews who were exiled in Babylon, the state of Judah having been defeated by Nebuchadnezzar. Six hundred years later, Christianity became the second Abrahamic religion, based on the early Bible lineage of Abraham's son by Sarah, Isaac. In contrast, the Arab religion of 600 CE was based on Abraham's first son, Ismail, "father of the Arabs," who, with Abraham, built the first Kaaba.

The oldest facts of a religion, such as whether Abraham existed in the second millennium BCE, become unverifiable. But there is an alternative "trace element" of number science within the Bible, which was transmitted and used at Mecca in the building of the Kaaba and *subsequent* creation of the Qur'an. As with Abraham's monotheism in the Bible, Islam shows signs that a similar school of advanced "angelic" influences were acting through the Prophet, as Muhammad indicated in his Qur'an.

The Bible's Abrahamic Innovations

The Bible's writers introduced sacred numbers (based on the octave) into the early Bible in a variety of different ways, most obviously through using the gematria of the new alphabets, such as Hebrew and Greek. These languages, based on a singular Phoenician invention, offered each phonetic character a letter-number correspondence. In ordinary life, separate number symbols such as 3, 5, and 9 were not then essential, while religious texts could and did use names to signify sacred numbers. For example, the first man, called Adam, is crucial to the Bible's descriptions of the relationship between humans and the one God who created them. Adam was given the number A.D.M = 45, which, when doubled to Sarah's age (90) at Isaac's birth is doubled again to Isaac's age

---

*"We believe in God and in what is revealed to us and what was revealed to Abraham and Ismail and Isaac and Jacob and to the tribes and in what was given to Moses and to Jesus and to the prophets from their Lord. We make no distinction between them. And to Him we submit." Qur'an, 3:83.

(explicitly given as 180 years) at his death. This leaves Isaac as an early and important octave, which, in tenths of a lunar month, forms intervals between the outer planets and lunar year (fig. 8.2).[1]

From the start of the Bible, the Abrahamic religion presented a cutting-edge version of an ancient model of harmonic order within the planetary world, in which the planets emerge harmonically from *one* God. While polytheistic religious texts also employ these techniques to express divinity alongside numerical tuning theory, their technique again probably dates from the first millennium. When applied to monotheism, the octave became a cornerstone for referencing harmonic patterns through a single number like Adam's. Similarly, using metrology, temple buildings could be based on the single limiting number required to quantify, as an integer, all of each building's internal structure.

It is wrong to automatically assume that Sumerian musicology of the third millennium BCE was similarly driven by the oneness of the octave, which was required in the Bible for a musicology expressing the unity of man with God: the harmonic model becomes more relevant in the context of the octave. The Abrahamic religions inherited their monotheistic octave from the, until then,

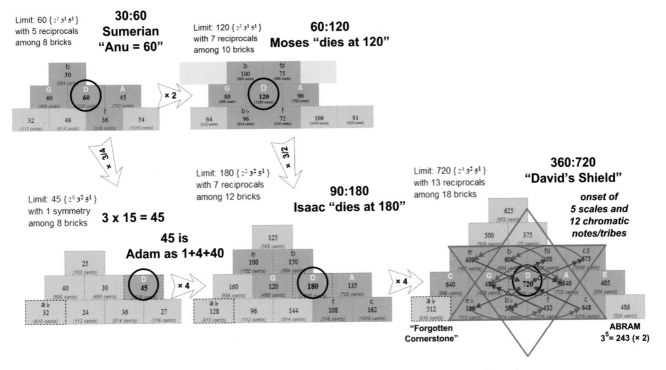

FIGURE 8.2. Abraham as Patriarch evolves the Children of Israel.

secret doctrine established at the dawn of history to characterize the planetary time ratios as within a given octave range given by the number of Adam, between 720 and 1440, so tied to the lunar month and year. The number 720 (6!) is the smallest possible number to describe an integer set of string lengths capable of expressing five modal scales in Just intonation, a smaller number limit than the 1728 limit required to form seven strings for the *single* scale of the Pythagorean heptatonic. The Kaaba expresses both 1440 and 1728 its inner and outer perimeters (see fig. 8.9, p. 192).

The Kaaba of Adam and Abraham

Muhammad extended the biblical story of Abraham in an extraordinary way, taking the birth and exile of his first son, Ismail, into the spiritual center of Arabia, Mecca. The story of Adam was also extended after his expulsion from the Garden of Eden. In this account, Adam was then a giant and was shrunk to a size appropriate for life on Earth. He treks from India through the Levant and down to what became Mecca. There he finds a dolmen made of gemstone megaliths,* from which the light of his soul illuminates the desert. The dolmen becomes scattered, but aspects or parts of it return in the narrative of Abraham and the Kaaba. When he builds a Kaaba near that spot, Ismail finds the cyclopean stones on which the present Kaaba still sits.[2]

The units on which the Earth, the Moon, and the planets are geocentrically harmonic are the lunar month and year, while the alignment chosen for all the later Kaaba foundations was the northerly maximum standstill of the moonrise every 18.618 years (the nodal cycle), a cycle sandwiched between the Saros eclipse period of 18.03 years—between near-identical eclipses—and the Metonic period of 19 years (fig. 8.3; see also fig. 2.11, the Third Triangle, p. 50).†

The present monument continues this alignment to the maximum Moon in the northeast, so that the Moon rises beside the eastern mountain during the Moon's orbital range (fig. 8.4). In each lunar orbit (of 27.3 days), the Moon also rises to the southeast half an orbit later, thus spanning the width of Mount Hira. This is where Muhammad's recitations began, and often continued, in a special cave. This mountain then *is* the range in angle of moonrise between

---

*This dolmen was created by God at the same time Adam was formed.
†These cycles, all involving numbers between 18 and 19, relate to the Equal Area Model and the Third Triangle, both discussed in chapter 2.

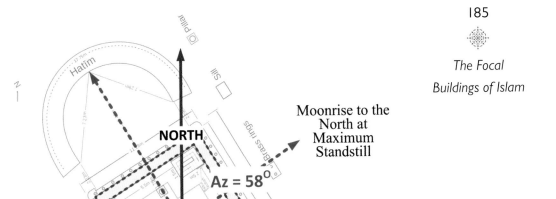

FIGURE 8.3. The alignment of the present Kaaba to the northerly maximum moonrise of the Moon and, at right angles, the rise of the bright southern star Canopus.

Illustration by Bojan Jankuloski for Wikimedia Commons.

FIGURE 8.4. Looking toward the maximum moonrise in the northeast over Mecca.

the Moon's most northerly and most southerly lunar risings as seen from the Kaaba at maximum standstill, and the placement and alignment of the Kaaba was chosen to achieve this.

The maximum moonrise (or moonset) is a rare but completely predictable event, occurring every 18.618 years (6800 days): a period appearing in the angelic mind because, when multiplied by π, one arrives at the perfect recurrence of the Sun every 33 years, as described in "The Squaring of the Circle in Area" in chapter 2 (p. 44). The 33-year cycle is linked, as a square, with the movement of the lunar nodes that cause eclipses and also with the orbit of the Earth as a year-circle. This solution to "squaring the circle" in area normalizes the Earth day with the nodal motion of the Moon.

The previous Kaaba buildings were said to have had the same rectangular foundations and were therefore similarly aligned. But Canopus, the important southern star and the second brightest in the sky, by 600 BCE rose at right angles to the maximum moonrise in the northeast. This was considered, esoterically, as a portent of the Prophet by the mystical Sufis but was outwardly expressed in the Islamic flag of the star and crescent—where the star is the morning star Venus with the crescent Moon, perhaps signifying when Muhammad and his army victoriously entered Medina.

There is a legend mentioning the actual dimensions of the Kaaba built by Abraham and Ismail. If the width of Muhammad's Kaaba (see the next section) was 11 units of 3 feet each, the foundation width for Abraham's building would

FIGURE 8.5. The Kaabas of Abraham's (*left*) and of Muhammad's time (*right*), both seen as one would look to the southwest.

*Adapted from ink illustrations of Mecca by Shaikh Tahir al-Kurdi
in Emil Esin,* Mecca the Blessed, Madinah the Radiant, *1963.*

also be 33 feet, numerically similar to the 33 years of the solar hero.* The ratio 33/18.618 will be the square root of π, due, as already stated, to the fact that a circle with a radius of 18.618 feet is of equal area as a square of side length 33 feet. That is, 1.772478247 squared is 3.141679, π accurate to one part in 36,000 (about 14.48 minutes in a year). A 32:37 triangle could be said to "have no roof and [be] open to the sky" because it represents what happens in the sky, something that is literally true of an observatory.

Ernest G. McClain provided a précis of Emil Esin's *Mecca the Blessed, Madinah the Radiant,* pages 22, 46, 73, and 133–44: "According to tradition, an earlier Ka'ba was trapezoidal: its four sides measured 20, 22, 32, and 37 ells, respectively, and the side which measured 22 ells was curved, and it had no roof."

> ***Geometrically,*** it so happens that the third side of a **32:37** right triangle is very nearly 18.618, the number of years between maximum standstills of the Moon, while the Kaaba is aligned to the northerly maximum moonrise, an alignment beside Mount Hira that recurs every 18.618 years. A **20**-by-**32** rectangle can become a trapezium with one of its diagonals equal to **37** as per fig. 8.6.
>
> ***Metrologically,*** if the width of Muhammad's Kaaba (see the next section) was, like the modern Kaaba, 11 units of 3 feet (33 feet), then 33 feet divided by 20 ells is 33/20 (1.65 feet), a cubit of 3/2 Saxon feet of 1.1 feet.

Figure 8.6 shows the present Kaaba as a gray rectangle over which the trapezium geometry and units in ell are shown. The 18.618-ell third side of the trapezium would allow a circle to be drawn, representing the nodal cycle as its diameter. The unit, an ell, can then be multiplied to define a square side length of 33 ells (54.45 feet), which is smaller than the circle's diameter (61.4394 feet). The ratio of the areas of the square to the circle will be π, while the ell then represents the solar year.

Another circular model can be seen in the Kaaba: the geodetic Equal Perimeter Model, then giving the Moon as 3 units (9 feet) in diameter to the

---

*Dividing 18.618 years into 33 feet gives 1.77247 feet per year, in Michell's *Ancient Metrology* a synodic Russian cubit.

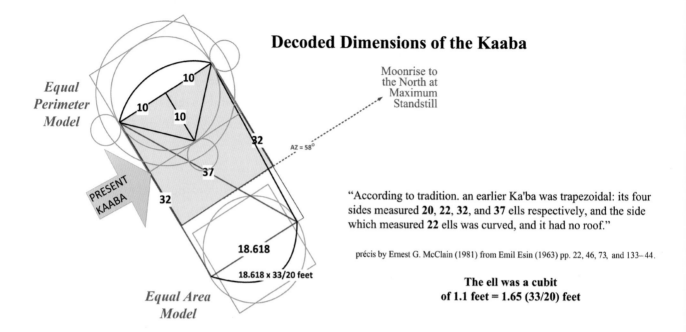

Decoded Dimensions of the Kaaba

Equal Perimeter Model

10
10
10
32
37
32

PRESENT KAABA

Moonrise to the North at Maximum Standstill

AZ = 58°

18.618

18.618 x 33/20 feet

Equal Area Model

"According to tradition. an earlier Ka'ba was trapezoidal: its four sides measured **20, 22, 32,** and **37** ells respectively, and the side which measured **22** ells was curved, and it had no roof."

précis by Ernest G. McClain (1981) from Emil Esin (1963) pp. 22, 46, 73, and 133–44.

The ell was a cubit of 1.1 feet = 1.65 (33/20) feet

FIGURE 8.6. The dimensions of Abraham's Kaaba interpreted to embrace the 18.618-year nodal period of the Moon's orbit (equal area) and a model of the Earth and the Moon (equal perimeter).

Earth's mean diameter being 11 units (33 feet), for the mean Earth radius of 3960 miles. When 33 feet are converted into inches, there are 396 inches in the width of the Kaaba, just one-tenth numerically (the scale then being 1 inch to 10 miles). Hence, the Kaaba design since Abraham was probably informed by this cosmological model with a diameter of 396 inches (see also the later section in this chapter, "The Model of the World in the Hatīm").

THE KAABA'S HARMONIC CODE

A recent plan of the Kaaba (fig. 8.7) indicates that its walls, by their odd proportionality, symbolized the numerical origins of musical harmony through the first six numbers {1 2 3 4 5 6}, with this sequence sometimes called the senarius, meaning "born out of six."

As already stated, in the cosmology of Plato's *Timaeus* the world was created using the rules of musical harmony in a scheme involving perfect fifths of

3/2, fourths of 4/3, and tones of 9/8, leaving "leftover" semitones of 256/243. This formed a rudimentary musical scale using only prime numbers {2 3} and their multiplication and division by themselves and by each other, a system often called tuning by fifths. The Kaaba incorporates another prime number, 5, called the human number by Plato, and this enables two more large intervals called thirds, the major third of 5/4 and the minor third of 6/5. Using prime number 5 enables more and better scales to be formed within an octave while filling the gap between the numbers 4 and 6 to then show all these large intervals as between the first six numbers, 1:2:3:4:5:6. This improved tuning system can be traced back to

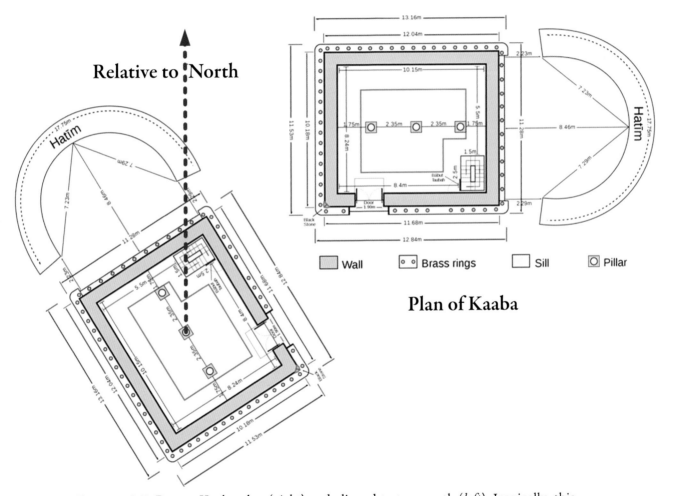

FIGURE 8.7. Recent Kaaba plan (*right*) and aligned to true north (*left*). Ironically, this plan was probably generated for the recent complete rebuilding of the Kaaba because otherwise surveying and archaeology are forbidden as intruding on the holy monument.

Illustration by Bojan Jankuloski for Wikimedia Commons.

the Sumerian tuning texts written in cuneiform circa 3000 BCE onward, though not necessarily in the context of scales within a single octave's tonic.

McClain wrote on the proportions of the Kaaba, a uniform spelling of the Kaaba here being imposed:

> What follows is an adventure in imagination which aims at grounding the Kaaba's proportions in the sacred sciences of earlier civilizations. Since Islam already claims a very great antiquity for the Kaaba, any explanation which appears plausible will merely have the effect of supporting the Islamic claim. My argument is based on the very interesting coincidence between the Kaaba's proportions—not its absolute measures—and those of the Temple of Poseidon in Plato's myth of Atlantis, a myth which has proved capable of very detailed explanation by the methods and metaphors of the Pythagoreans. The temple of Poseidon is described as involving the ratios "6:3 plethra (full)" meaning 6:5:4:3. The Kaaba's proportions are precisely these:

	Width	Length	Height
Measures (in meters)	10	12	16
Proportions (in smallest integers)	5 :	6	
		3 :	4

We can say that the Kaaba "embodies" the ratios 3:4:5:6, the numbers 3 and 6 being an "octave identity" in Pythagorean harmonic theory. We are investigating not the visible geometry of the Kaaba but the proportions which have their meaning in the invisible sonar implications of a number theory which the whole ancient world shared in common for several millennia before Muhammad.

Muhammad, as an "unlearned" man, has no share in this ancient science; he is merely the agent—as he conceived himself to be—by which ancient knowledge came to new life in Arabia. Before embarking on our new study of the Kaaba, let us first notice the role which number plays in the Quran. My whole study would be invalidated—at least for a Muslim—if the Quran itself did not virtually ignore geometry and laud, instead, number and proportion.[3]

My own measures for the monument (derived from the previously mentioned plan, figure 8.7) *seem* different in giving a width of 11 yards rather than 10 meters, but 11 yards are 10.0584 meters, while McClain's figures (after those of the Swiss German author Titus Burckhardt) were only approximate, being in meters. My length is 39 feet, which is 11.89 meters, that is, nearly 12.[4]

My own geometrical proposal embraces the whole senarius through noting how the Kaaba was built as the outer wall of a set of nested rectangles, as illustrated in figure 8.8.

The walls of the Kaaba are as if constructed by blocks one-yard square

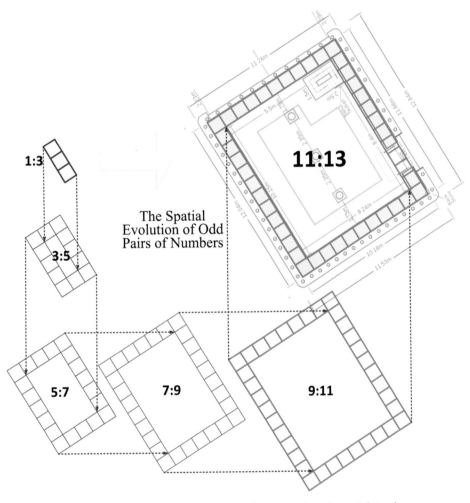

FIGURE 8.8. The emergence of proportionality within the
Kaaba's grid of 143 square yards.

Underlay illustration by Bojan Jankuloski for Wikimedia Commons.

(fig. 8.8). The inside floor area would then be 9 × 11 = 99 square yards, the number symbolic of the ninety-nine names of Allah. When seen as a set of nested rectangles, the sequence 13:11:9:7:5:3:1 arises, in which 7 is the middle term and the sum is 49, or 7 squared (see fig. 8.11, p. 194), 7 being the number that terminates the senarius.

In figure 8.9, the walls of the Kaaba are again seen to be meaningful, in inches as well as feet and yards. The outer perimeter is 12^3 = 1728 inches, symbolic as the "head" number specified for Near Eastern arks, while the inner perimeter is 1440 inches, symbolic of Adam when his letter numbers of 1.4.40 are placed in decimal position notation.[5]

Obviously, sacred buildings are somewhat equivalent to arks, while the biblical figure of Adam, further developed in the Prophet's new narrative, has the original Kaaba as a dolmen filled with light that was created by God near Mecca when Adam was made. Abraham and his first son, Ismail, built the earliest version made by human hands, and Muhammad himself participated in a major rebuilding of

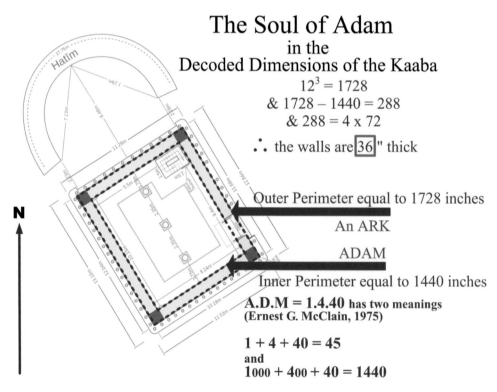

FIGURE 8.9. Inner and outer perimeter lengths of the Kaaba in inches.

Underlay illustration by Bojan Jankuloski for Wikimedia Commons.

the Kaaba after flood damage, which could have been designed to the present Kaaba's proportionality, with later works then based on that footprint.

It is the perimeter lengths of the Kaaba's nested rectangles (fig. 8.10) that tell the story of the senarius.

TABLE 8.1
THE OUTER PERIMETERS OF NESTED RECTANGLES
IN THE KAABA THAT INDICATE THE SENARIUS

Half Perimeter	Perimeter	Senarius	Interval
3 + 1 = 4	4 × 2 = 8	8/8 = 1	Unison (1:1)
5 + 3 = 8	8 × 2 = 16	16/8 = 2	Octave (1:2)
7 + 5 = 12	12 × 2 = 24	24/8 = 3	Fifth (2:3)
9 + 7 = 16	16 × 2 = 32	32/8 = 4	Fourth (3:4)
11 + 9 = 20	20 × 2 = 40	40/8 = 5	Major Third (4:5)
13 + 11 = 24	24 × 2 = 48	48/8 = 6	Minor Third (5:6)

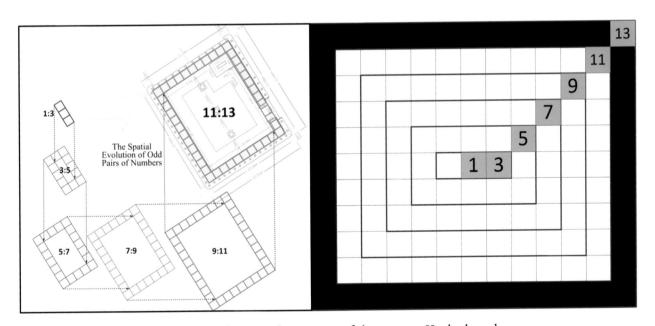

FIGURE 8.10. The nested structure of the present Kaaba based on a single unit 3 feet or 36 inches square.

Underlay illustration for image on the left by Bojan Jankuloski for Wikimedia Commons.

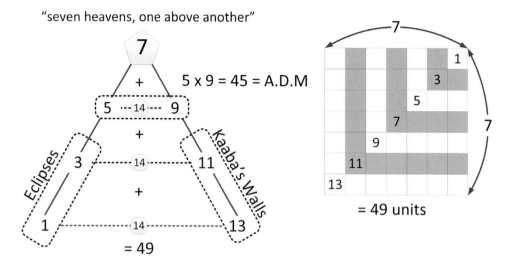

FIGURE 8.11. The cumulative properties of gnomonic odd-number pairs when squaring their mean member, in this case, 7, summing to 49.

This reveals special properties that are implicit in the Kaaba and can be found in the early odd numbers as a series. Such numbers are now called gnomonic since, cumulatively, odd numbers sum to the square they would define on the diagonal, as in figure 8.11. Any rectangular structure made of one-unit-square bricks would, if made 13 long and 11 wide, define an inner space within which all the odd rectangles before it would fit down to a wall 3 by 1 units, as per figure 8.10.

It becomes obvious what the height of the Kaaba should be in order to continue the series of odd numbers:

$15 + 13 = 28 \times 2 = 56$, which is 7×8 yards (56 yards in perimeter).

This gives the senarius as the floorplan of the Kaaba, broken by the prime number 7 (its height), which does not participate directly in musical harmony except structurally in the seven distinct notes of the octave, the seven types of diatonic scale, and so on. This dimension of 7 presses upward, creating a third dimension indicative of a gateway into the sky and heavens through the building's roof, dimensions invisible to the ordinary eye. As such the Kaaba resembles the monolith used in Stanley Kubrick's 1970 film *2001: A Space Odyssey*, which was based on a novel by Arthur C. Clarke. In the novel:

when the Monolith [on the Moon] is excavated and examined, it is found to be a black cuboid whose sides extend in the precise ratio of 1:4:9 or $(1^2:2^2:3^2)$. ... Clarke suggests that this sequence or ratio extends past the three known spatial dimensions into much higher dimensions.[6]

As a thought experiment, the dimensionality can be increased to five in all to yield numerical limits that are very appropriate to the longer-term behavior of the Moon (the Saros period of 19 eclipse years and Metonic period of 19 solar years) and the Islamic maxim that 19 is the signature number of Allah.

It is profoundly significant that the two sacred numbers, 1440 and 1728, should naturally emerge from the three-square geometry that encapsulated the meaning of the monument's orientation when 36 inches is taken as the unit length and both sides of the rectangle are repeatedly lengthened by two units. Most ancient temples employed grid designs, and most ancient mythology seems to have found a use for 1440 and 1728 in some of their greatest stories, but the Kaaba links, in an original and probably unique way, these mythological numbers to the astronomical facts through the numerical geometry of its design and its orientation of the Moon's maximum standstill.* After quantifying these possible higher dimensions of the monument, we can the return to the model of the Earth built within it and the semicircular Hatīm.

DIMENSIONS BEYOND THE KAABA

A four-dimensional cube can be visualized as a cube within a cube, and here the new odd number is 15 + 2 = 17, the prime number seen in the periodicity of the lunar nodal period of 6800 days (18.618 years) and in the nodal megalithic rod and yard (of 6.8 and 2.72 feet). This can be deconstructed as 17 × 400 days between one maximum (or minimum) standstill of the Moon and the next, the exact cycle to which the Kaaba has been aligned since the megalithic-style monument was built by Abraham and Isaac.

$$15 + 17 = 32 \times 2 = 64/8 = 8 \text{ yards} = 2304 \text{ inches}$$

---

*"The term 'lunar standstill' was apparently coined by Alexander Thom, in his 1971 book *Megalithic Lunar Observatories* (Oxford University Press). It is analogous to the term 'solstice'; in neither case does the Moon or the Sun actually stand still." Vincent, "Major 'Lunar Standstill.'"

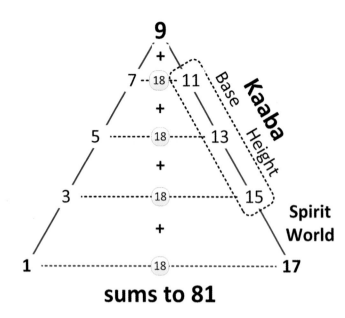

FIGURE 8.12. The spirit world of the Sufi is higher than that of bodies but less than the spiritual world. Here is the system defined by the number 9.

When 17 is the limit, then 9 is the odd number in the middle, and all the odd numbers add up to the square of 9, which is 81 (fig. 8.12).

For an Islamic interpretation, one can reference the technical language of Sufism:

1. The World of Bodies (*ālam-i-ajsām*): the Kaaba as a physical body
2. The World of Spirits (*ālam-i-arvāh*) as the fourth dimension, linked in eternity to time.

In harmonic terms, the number 81 is the fourth power of 3, which is associated with the planet Venus and the ancient goddess Inanna/Ishtar, while 81/80 is the synodic comma of Adam's Just intonation, a ratio found between the lunar and sidereal months in the harmonic model.

On the next page, figure 8.13 explores the further increase to lengths 19 and 21. The Spiritual World (*ālam-i-imkān*) can be imagined between 17 and 19:

$$17 + 19 = 36 \times 2 = 72/8 = 9 \text{ yards} = 2592 \text{ inches.}$$

The Sufic Unfathomable (Lahut) exists between 19 and 21:

$$19 + 21 = 40 \times 2 = 80/8 = 10 \text{ yards} = 2880 \text{ inches.}$$

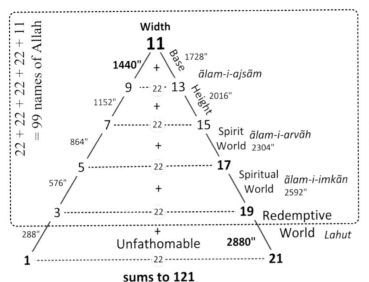

FIGURE 8.13.
The spiritual
world of the Sufi

These later rectangular perimeters, in numbers of inches (2880, 2592, 2304, and 1728) are compatible with the harmonic matrices of Ernest G. McClain. Thus, 1440 is an octave below 2880, and so the "holy mountain" for 2880 is highly relevant, especially since (a) 1440 is the greater number of Adam when interpreted in position notation, and (b) 2880 is the arrival of the eclipse year in the harmonic model—the Unfathomable and eclipsing of the existence of the Sun or the Moon (fig. 8.14, p. 198).

If the perimeter distance of a 19 by 21 yard rectangle is 2880 inches, one can ask:

- Why is this beyond Adam and apparently at the head of the spiritual world (where one would place Allah), and
- What difference does 2880 bring to the holy mountain of 1440?

The second question can be answered if, in the planetary matrix, one sees that the two tritones opposite D, which is the lower limit of 1440 and the upper limit of 2880,* become "symmetrical twins" (Saturn equals 2048 and the other, the esoteric Centaur asteroids, equal 2025) plus there is one new brick on top of 1875, which is then the eclipse year of 346 days in units of 1/80th of the lunar month. These two phenomena, the musical tritone to the tonic D

---

*See also figure 5.13 for the Olmec-Maya significance of 2880.

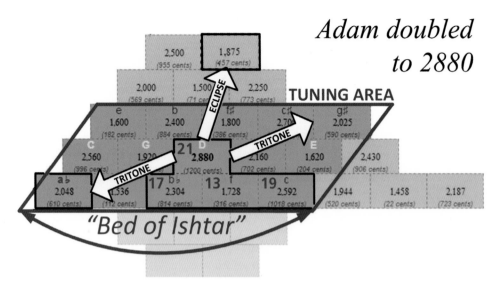

FIGURE 8.14. Adam doubled to 2880 brings a
new relationship to the Moon's eclipse.

and the astronomical period between possible eclipses on the same node, thus
arrive together in the matrix of planetary harmony "above" Adam as 1440. In
an eclipse, a light body is briefly extinguished, and in Sufism, a lower aspect of
selfhood is annihilated (*fana*) before reaching each station (*maqam*) of being.

THE MODEL OF THE WORLD IN THE HATĪM

Returning from the implied but invisible extensions of the Kaaba beyond its
floor plan, it is interesting how the width of 396 inches invites the now familiar
model of the mean Earth and the Moon to be formed (fig. 8.6, p. 188), which
then embraces the Hatīm with higher dimensions, numerically to 19, as congruent with that model.

In figure 8.15, opposite Moon circles would define a diameter of
17 (11 + 3 + 3) so that, in yards, the model of the Earth is in its native units
(here, for example, the diameter of the Moon circle is 3 yards, and the sublunary
world is 14 yards) and so could be said to belong as formed in the angelic mind.*

---

*We see this number 17 as a painted brass plate (fig. 7.10, p. 158), on the roofline circumference
within van Orley's *Annunciation,* linked to the Glastonbury Old Church.

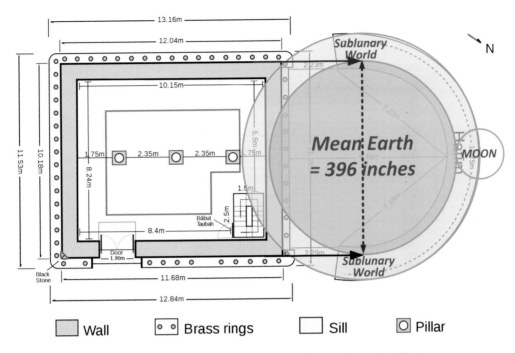

FIGURE 8.15. The Hatīm's and the Kaaba's sides of 396 inches conform to the 11/3 model of the Earth and the Moon.

Underlay illustration by Bojan Jankuloski for Wikimedia Commons.

In figure 8.16 (p. 200), the inner diameter of the Hatīm can be seen to be 10 yards (counting the green squares) while, in figure 8.15, the inner circle of the Equal Perimeter Model is diameter 11 yards. In figure 8.16, the seven right-most (purple) grid squares (each a yard) show the Hatīm's outer diameter is 14 yards, making the outer circumference, if a complete circle, 44 yards of 36 inches = 1584 inches which is 11/10 of 1440 inches while 1728 is 12/11 of 1584 inches. The Hatīm's outer circumference, signifying the sublunary world is therefore the arithmetic mean between the outer and inner circumferences of the Kaaba itself.

The circumference of 1584 is exactly half the 3168 units that are normal to cosmological models found between Europe and the Near East since 2500 BCE (see chapter 2), remembering that John Michell found 3168 had an ancient sacred function of delineating a sacred space. In the case of the Kaaba and its size, the Hatīm appears to be an original feature of *Abraham's building*, realized at the appropriate scale for that building. In figure 8.6 (p. 188), one can see that the Equal Perimeter Model was probably further northwest and the limit of its

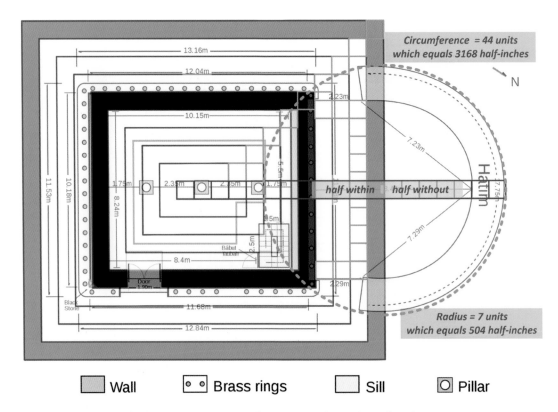

| | Wall | | Brass rings | | Sill | | Pillar |

FIGURE 8.16. The Hatīm is a significant metrological circle relating to the sublunary world. This resembles the domes mentioned earlier in that the inside (the Hatīm) represents the sublunary world and is half-inside and half-outside the unfathomable world limit of a 21 yard rectangular side length.

Underlay illustration by Bojan Jankuloski for Wikimedia Commons.

center 21/2 yards from the present Kaaba's center.* As should always be the case, the outer radius of the Hatīm is a number involving 504: the radius of the mean Earth (5.5), and lunar radius (1.5) added together, as $7 \times 36 \times 2 = 504$ inches. The other main components of this cosmological model can be added in figure 8.17.

The components and units fit the lunar diameter perfectly, reaching the limit of 17 as $(2 \times 5.5 + 2 \times 3 = 17)$. The mean Earth is 396 half-inches, which is numerically one-tenth of the mean Earth in miles. This gives the scale of the model relative to the Earth of one-half inch to 10 miles (remembering that

---

*If the older building was 32 ells long of 1.65 feet (52.8 feet), then a 21 yard side length (53 feet) for the extended grid was just 1/5th (0.2) feet longer than the 32-ell-long building.

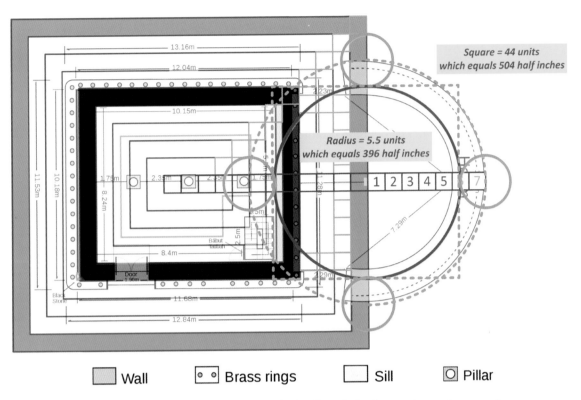

Square = 44 units
which equals 504 half inches

Radius = 5.5 units
which equals 396 half inches

| 1 | 2 | 3 | 4 | 5 | | 7 |

Wall   Brass rings   Sill   Pillar

FIGURE 8.17. The mean Earth circle (*red*) and the lunar circumferences in quadrature and the square of equal circumference to the outer circle (*blue*).

Underlay illustration by Bojan Jankuloski for Wikimedia Commons.

Adam was shrunk). In summary, the Kaaba can be seen to equal the square of equal perimeter to an outer circle that denotes a sacred space (as the 3168 half inches of the Hatīm), using that model's native units of the yard based upon a nested set of odd rectangles.

THE DOME OF THE ROCK

When Islam inherited the site for the Temple Mount from Byzantium in the seventh century CE, the Dome of the Rock was built above the exposed expanse of bedrock from which Mohammad describes flying toward heaven on his Night of Power. The dome of this building (692 CE) above that rock has a diameter that is half that of an out-circle drawn *within* the octagonal internal pillars that formed an observational ambulatory.

The ambulatory's inner diameter was 39.6 cubits of 1.65 feet, with the ell originally defining the Kaaba, and the inner diameter of the outer octagonal walls was 79.2 ells, equaling 130.68 feet. This numerically resembles the mean Earth diameter, which, divided by 11 and multiplied by 14, gives 100.8 ells (166.32 feet) as the sublunar boundary in the Equal Perimeter Model, as shown by the out-circle of the octagonal building's outer walls. The Dome of the Rock is therefore a perfect example of the Equal Perimeter Model in the ell units of 1.65 feet, used to define Abraham's building (fig. 8.18).

The Moon, in 3/11 proportion to the Earth, would then be 21.6 ells, or 35.64 feet, somewhat smaller than the Dome of the Chain nearby, which is

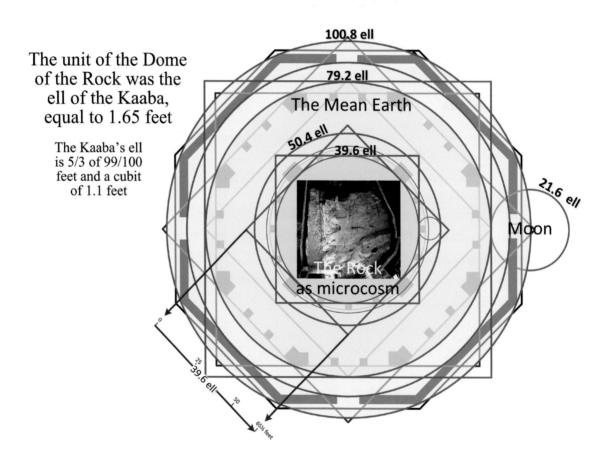

FIGURE 8.18. The Dome of the Rock with the combined model, achieved using the regular octagon's wall thickness.

FIGURE 8.19. Photo of the Dome of the Rock and, before it,
the Dome of the Chain.

Photo by Andrew Shiva for Wikimedia Commons.

26.4 ells, numerically the size of the Old Church at Glastonbury in feet and so
1.65 times larger than the Old Church. This smaller dome was built to demonstrate the overall concept to Caliph Abd al-Malik, who had plans for the

Dome of the Rock to become a competing focus for Islam to Mecca, its inset dome having a diameter half of 26.4 ells. The chain in question was a metrological one, of two circles, one inside the other, whose combined circumference equals a "chain" of 66 of the feet (99/100 feet) chosen by its architects for the Dome of the Rock,* 5/3 of Abraham's ell.

Michell, in adopting the Classical notion that humans were a microcosm of the macrocosm, found a number of inner circles that were half of the outer circle of a monument. The Rock perhaps signified the human couched within the world. The Rock has been encircled by more than one story involving a human drama: Abraham's reprieve from sacrificing Isaac, the nearby Temple of Solomon and later Jewish replacements, the ministry of Jesus, and Muhammad's night journey at Jerusalem, the latter arranged by the archangel Gabriel.

The Dome represented the fact that the octagon's shape could enable the combined Equal Perimeter and Equal Area Models if the equal area circle just touched the inside centers of the outer walls as their in-circle, while the equal perimeter circle (14) just touched its outer vertices. This required the wall to have a specific thickness of 4 × 99/100 = 3.96 feet, or 2.4 ells. The shared square of the combined Equal Area and Equal Perimeter Models is then 33 of these units, and the units then form the 18.618-unit radius of the circle equal in area to that square. The whole edifice was therefore a *tour de force* in both metrology and the models of the Earth and its Moon, expressed in ells rather than in feet or the other units found in earlier chapters.

Since these models have evidently been present for millennia, their purpose remains quite obscure as to why they were built long after the tradition from which they arose had clearly disappeared from the outer life of subsequent civilizations. Religion now is obviously tied to the notion of sacred space but without any concrete notion as to why and how this was done. This matter of purpose requires a will to purpose it, and the absence of a purpose for the design of these monuments leads us to extra dimensions, apart from existence and eternity, to explain it: a *dimension of Will*.

---

*"Unwilling to commit himself to the construction of the Dome of the Rock without first examining a model of the proposed structure, Caliph Abd al-Malik commissioned his architects to erect a scaled-down version of the Dome. That structure, which is known as the Dome of the Chain, stands today in the shadow of its full-sized neighbour." Landay, *Dome of the Rock,* 1972.

9
Cosmological Numeracy

WE CANNOT KNOW WHICH SPIRITUAL TEACHERS of the twentieth century will remain relevant in the twenty-first to help us overcome our perceived lack of cosmological purpose. One enduring influence for me has been G. I. Gurdjieff, who identified this lack of purpose with a great question for himself:

> What is in general the sense and meaning of man's existence?[1]

His question grows in significance in light of the unusual numerical circumstances of the Earth and hence of our lives.

GURDJIEFF AND BENNETT

Gurdjieff's first cosmological ideas, presented in prerevolutionary Russia, were based on a scheme of world numbers that corresponded to levels of scale within the universe: (1) the Absolute (all), (3) all worlds (galaxies), (6) all suns (Milky Way), (12) the Sun, (24) the planets, (48) the Earth, and (96) the Moon and the Absolute (nothing)—all of this contained in a Ray of Creation labeled as if it were a {*do re mi fa sol la si do*} octave (fig. 9.1, *left*). These were preserved in *In Search of the Miraculous* (*In Search* hereafter), a collection of notes taken of his talks, which was edited by P. D. Ouspensky and eventually published in 1950, shortly after Gurdjieff's death.

Gurdjieff's use of number reminds us of Pythagoras, who proposed that the planetary cosmos was harmonic, yet Gurdjieff's proposal was a much wider one in which *all processes* in the universe were harmonic, not just the planetary

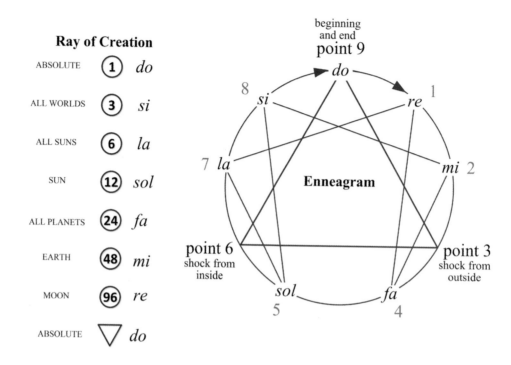

FIGURE 9.1. The Ray of Creation (*left*) and the Enneagram (*right*),
which resembles the tone circle for an octave.

time cycles seen from the Earth. Within his Ray of Creation, our Sun was the
ascending dominant tone, strangely called *sol* in our solfège notation {*do re me
fa sol la si do*}. Gurdjieff also had a circular diagram called the Enneagram, with
nine points, again with solfège notation, *do* being at the top, then with numbers and note classes rising clockwise, with *sol* at point 5. These systems were
not as simple as figure 9.1 implies since, in his presentations, the Enneagram
appeared to apply at many different levels of scale but in essence was an important way to explain how octave processes need to successfully progress, by need-
ing help from outside at point 3 and help from within at point 6. Unlike a
musical octave, Gurdjieff's octaves belonged to a "science of vibrations" that was
alchemical, so acoustic music shares the outer form of cosmic transformations
(the octave) but lacks something necessary in the cosmic world to counteract
the increase in entropy within time itself, later called the Merciless Heropass by
Gurdjieff perhaps because only the heroic can overcome time.

After a serious car accident in 1924, Gurdjieff instead decided to pass on his formerly theoretical teachings by writing an epic tale. *Beelzebub's Tales to His Grandson* (*Tales* hereafter) was a cosmological description of the history of the Earth, set in the epic style of oral tradition. The directness of his Russian diagrams and lecture format was replaced by a traditional epic—something familiar to him since his father had been an exponent of this tradition, locally called an *ashokh.** I believe the Enneagram and the Ray of Creation were still latent within his new, textual approach, but its unusual style of writing forced readers to make better efforts in understanding a then more fully developed version than "fragments from an unknown teaching," which was the sub-title of Ouspensky's *In Search*. Departing from the theoretical harmonists of Pythagoreanism, Gurdjieff's influences came from further East, and the Enneagram of *In Search* was transmogrified into the "law of ninefoldness" of *Tales*, then called Heptaparaparshinokh.

In *Tales,* all material processes were to be governed by two fundamental "laws," one of seven parts and the other of three parts. These independent laws were now combined by God, who made them the primary cosmic law from which a relatively independent universe would then arise. The octave of seven parts and the triad of three parts (affirming-denying-reconciling) came to govern *every process* in the universe as to its possible transformations within a law of ninefoldness. This divine design had the *express purpose* of making God's own existence independent of time and entropy through the mechanism at point 3 in the Enneagram (the third interval of his Heptaparaparshinokh), which would ensure that all beings would both be fed and feed one another, according to a law of *reciprocal maintenance*, his Trogoautoegocrat.

It is possible to see this, in modern terms, as the harmonious complexity of the worlds due to their interdependence, intuitively attractive because at first sight God can withdraw to our big bang and let the universe do its thing: The notion that God should *not* control His universe is very different from the idea of God most religious people believe in, while to modern science no God is required anyway in a cybernetic universe of self-organizing systems. Gurdjieff's earlier system had reduced all cosmic systems to a single regime of worlds that were numbered using relatively small integer numbers, making his use of number seem unlikely.

---

*Someone who is "trained from a young age to remember and recite epic tales, in the manner of the ancient world." Gurdjieff, *Meetings with Remarkable Men*, 32.

In *Tales*, "World creation and maintenance," governed by a ninefold law, was the form of the Universal Will that is organizing *at all levels* according to a science of vibrations, governed by numbers and related to octaves. Such a universe would be scale-invariant and self-similar, at all levels of detail, to the overall form. However, looking into the detail given by *In Search,* the law of ninefoldness is purely planetary, emanating from the Sun (World 12) as a side octave of Life.

In Russia, Gurdjieff was concerned with moving human evolution out of its current state of sleep and delusion. Such an evolution would involve the manufacture of higher substances out of lower substances, resembling the work of the alchemist. The more sophisticated cosmology of *Tales* now sits alongside details found in *In Search* and the psychological and physical practices taught to Gurdjieff's groups of pupils. The three aspects of three-brained beings (thinking, moving, and feeling) can be seen as an effective law of three for the microcosm, the law that was added to the law of seven, making possible a "waking up" of consciousness through self-observation, self-remembering, attention to three kinds of food, including breath and impressions, having an aim, and so on.

The multiple perspectives of Gurdjieff's earlier and later ideas can today interact in useful ways so that the formal lectures of *In Search* complement *Tales* and his other books. Some demand the purity of his teaching or his writing, while others have brought yet other influences to bear, and, indeed, one might argue this was Gurdjieff's own technique, which enabled him to work with ideas in the first place—from "outside." For in the path of Will* there are no absolute prescriptions but, rather, little-explored laws involving numbers and harmony, in the domain of Will.

The Development of Egoism

The system of time seen from Earth, which expresses musical intervals between celestial parts, might well relate the harmonic model to Gurdjieff's cosmology. For example, the emergence of the human essence class coincided with the emergence of the harmonic intervals between the lunar year and the outer planets.

This onset of harmonization can be graphed using equations that can model historical changes in the Moon's orbit (fig. 9.2). In chapter 5 and elsewhere I have

---

*In Ouspensky, 1950, pages 41–42: Will is equated with the Theosophical notion of a causal or divine body in which there are many contradictory "wills" created by desires, hence the need for a path toward unifying these.

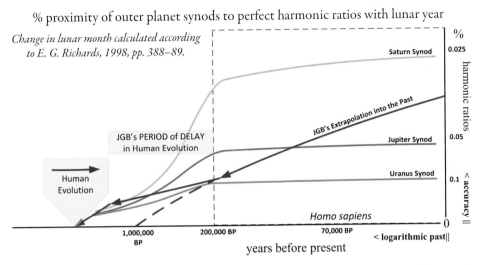

% proximity of outer planet synods to perfect harmonic ratios with lunar year

FIGURE 9.2. The emergence of modern humans alongside the onset of harmonic intervals between a lunar year, which is getting longer, and the fixed synods of the outer planets. J. G. Bennett's timeline* for the delayed evolution of humans due to cosmic circumstances is underlaid (*as a black line*) starting with truly human evolution *(yellow boxes)* as our evolution appears linked with the increasing resonance of the moon with the outer planets.

shown how the lunar month and lunar year are the common denominator for harmonic resonance with the other planets. Also shown in figure 9.2 is J. G. Bennett's estimate for how long humanity was *held back in its development* by the imposition, described in *Tales,*[2] of the organ called the Kundabuffer, through which the proper functioning of human conscience and reason was rendered blind to the objective reality of their circumstances. We are told in *Tales* that this was necessary to harmonize the accidentally created Moon. Figure 9.3 shows the arising of *Homo sapiens sapiens* with the onset of planetary harmony.

A million years ago, before that harmonization of the Moon to the outer planets, humans had the necessary brain development and ability to use tools, but their evolution had stalled. When the organ was removed, *Tales* asserts that its habitual propensities lived on within the human through habit, and in these habits Gurdjieff found an origin for human egoism, the self-love and greed of today's *Homo sapiens sapiens.*

Gurdjieff's plot, involving the creation of the Moon by a cosmic accident

---

*Bennett, *Dramatic Universe,* vol. 4, figure 46.2, section 17.46.3, pp. 220–26.

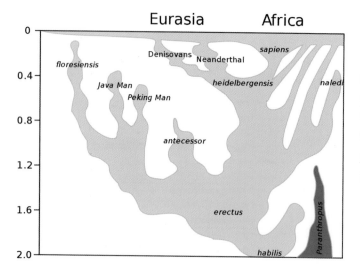

FIGURE 9.3. Chart of the phylogeny of the *Homo* genus with time past descending from the top in millions of years.

Image by Dbachmann for Wikimedia Commons

and the need for that body to become harmonized with the greater cosmic order, corresponds well with recent science's contention that the Moon was created through a planetary collision. It seems the Moon has been influential in human development, and the Harmonic Model, of the Moon harmonized with the outer planets, is a manifestation of this. His story therefore suggests that, the organ that was *placed in mankind to make the Moon's orbit harmonious* coincided with the emergence of *Homo sapiens sapiens*.

This conforms to my thesis that megalithic astronomy was an angelic intervention once the minds of *Homo sapiens sapiens* were ready to develop minds by the end of the Stone Age. If so, removal of the organ around 200,000 years ago, after the harmonization of the Moon, could then develop human species until, after the last ice age, megalithic astronomy prospered 5000 to 7000 years ago, long before metal instruments, lenses, telescopes, or arithmetic methods. Civilization soon arose that employed the new mental faculties to develop new tools, and the Neolithic revolution was able to support the cities known to history.

Given that Gurdjieff mentions the creation of the Moon, the imposition of the organ Kundabuffer, and the lingering psychological problems of its removal, his story probably drew on a special source, whom Bennett called the Masters of Wisdom, or even from the subconscious higher centers using hypnosis.*

---

*One can read a semiautobiographical account in *Tales,* 1132–37, where Beelzebub describes using a medium to locate the missing halves of the *Boolmansharo,* an Atlantean tablet describing "The Affirming and Denying Influences on Man."

J. G. Bennett developed a protoscience for will called Systematics, in which the numbers {1–12} form structures of will in a new framework dimension called Hyparxis. This invariance for numbers is like the phenomenon of harmony, which is held in the framework dimension called Eternity. Systematics is a different number invariance held within Hyparxis. This suggests that the role of numbers within the three different framework dimensions is similarly developed as (a) numbers becoming measurements (Existence), (b) numbers forming the musical intervals (Eternity), and (c) numbers becoming different but complementary aspects within situations (Hyparxis), within the Universal Will. The three framework dimensions can be seen most simply as the sides of the equilateral triangle within the Enneagram, whose points are often given as Being, Function, and Will, as in figure 9.4.

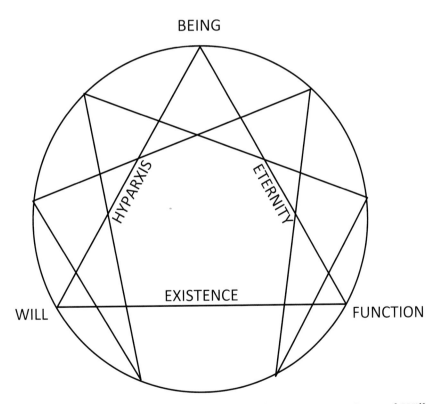

FIGURE 9.4. The three points, conventionally Function, Being and Will, can be seen as interconnected by the framework dimensions Eternity, Existence, and Hyparxis.

Each of the three dimensions has three aspects, called Function, Being, and Will, which Bennett called the Fundamental Triad of Experience. "Function, Being and Will [are] the ultimate modes of all possible experience. In this triad, will is affirming, function denying and being reconciling."[3]

This can be shown as table 9.1, with nine permutations for which terms can be found, as seen from the Earth.*

TABLE 9.1
THE THREE FRAMEWORK DIMENSIONS WITH THE THREE FUNDAMENTAL ASPECTS RELATIVE TO THE GEOCENTRIC WORLD*

Function	Being	Will	Aspect Dimension
Adam	Heavenly Host	Universal Will	**Hyparxis**
Biosphere	Sun and Moon	World Laws	**Eternity**
Earth	Planets	Stars	**Existence**

*This table has elements whose meaning will mature when more of Gurdjieff's and Bennett's ideas are explored.

The Sun and the Moon are the (geocentric) Being aspects of Eternity. Everything arises from the Universal Will, and the will aspect of Eternity are the world laws. The functional aspect of Eternity is the biosphere, a dynamic set of interrelated patterns in Eternity that are found in the biosphere. The cosmic laws have led to the stars and our Sun, the will aspect of Existence.

A new planetary world emerges around the Sun as the Being aspect of Existence, while in Existence, Function is the Earth itself. In the solar system, the being aspect of Hyparxis is the Heavenly Host, responsible for the Eternity of the planetary world and the creation of the biosphere through the central agency of the Moon illuminated by the Sun (the lunar year and month).

Through long study and direct experience, Bennett derived his intellectual scheme from Gurdjieff's worldview and the then current state of knowledge, a scheme which I have greatly abbreviated here. Bennett's legacy is a rare opportunity to understand why Gurdjieff's teachings advocated a work on ideas alongside a work on oneself. Bennett forged a new description of how human beings

---

*This technique is like Ouspensky's division of the three centers as each having three parts of the same nature: moving, emotional, and intellectual.

could interact with higher intelligence and, through this, for example, contact the *form* of history and understand its purpose. He wrote, in his preface to *The Dramatic Universe,* volume 4:

> The main thesis and its corollaries are unfashionable, chiefly because they appear to be a return to a doctrine of providential history that disregards the Laws of Nature. This is certainly not true for the scheme I have put forward. The operation of human Intelligence does not violate the Laws of Nature, even when it results in changes in the course of events. The Higher Intelligences that I have postulated must work in the same way, though on a far greater scale of time and space. If this postulate proves to give us a satisfying account of the traces of the past that we call "history," it should also guide us in our expectations of the future and in framing our decisions for action within the present moment.[4]

To understand history in terms of Being and Will inevitably involves understanding intelligences higher than our own, different from us in that they operate as long-term *influences* on existential situations—their present moment being greater than ours. Invisible influences were carefully eliminated during the creation of modern science, but Bennett had to reintroduce them, not as animistic forces but as influential forces, enabling human minds to achieve more intelligent outcomes.

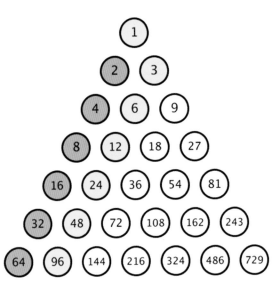

FIGURE 9.5. Gurdjieff's Creation of Worlds {1 3 6 12 24 48 96} as an extended Tetractys, through reconciliation (3) (*yellow*) of the denying force (2) (*red*), by the will of God (1). This may usefully be compared to figure 9.1 (*left*).

Three and the Partition of Will

It is not clear in figure 9.5 why the world numbers {1 3 6 12 24 48 96} of Gurdjieff's cosmology consist of the number 3 multiplied by successive powers of 2. Are these mere labels, or are these numbers reflecting something happening within and between the worlds? The number 2 expresses doubling, and 3 expresses the cosmological step from 2 to 3 at the top of the Tetractys. Whatever created the worlds enumerated by Gurdjieff was 1, the Universal Will that created the universe through this triad of expansion {1 2 3} therefore, all of Gurdjieff's worlds have the prime number 3 as their root number, then doubling between the worlds.

Bennett explains the triad as three successive actions.

> Progress beyond the dyad—but not its annihilation—is by entering into action. Now all activity is initiated by acts of will. We distinguish three moments in the realization of events. First, the **act**, which establishes the dynamism by the contact of impulses. Second, the **action**, which sets a process in train by the relatedness of impulses. Third, the **activity**, the "ongoing." To give a rough mental picture of the term characters: affirmation is the impulse behind commitment; receptivity is the impulse that opens up a field of action; reconciliation is the impulse that enables the dynamism "to be." In concrete situations the impulses interpenetrate and blend with each other.[5] (Emphasis added.)

One of Bennett's great insights was to see that lower worlds result from partitioning the higher world into two near-identical portions, one more pure and the second more tainted by existentiality. This can explain why Gurdjieff's lower worlds are doubled in the number of their laws by this partitioning—an idea similar to but different from doubling within a musical octave. Something unified in the higher world is transformed into a lower world, one true to the higher world but blended with an inferior copy, as to its laws. Thus, in World 3, three forces are differentiated but not connected:

1. Affirming,
2. Denying, and
3. Reconciling.

These are then blended through permutation into the six laws of World 6:

TABLE 9.2
THE SIX PERMUTATIONS OF THREE FORCES
INTO THE SIX TRIADS OF WORLD 6

3-1-2	Order	Part 1. Elaboration of the Universe
1-2-3	Involution	(see fig. 9.6)
2-3-1	Identity	
1-3-2	Interaction	Part 2. Fulfillment of the Universal Will
3-2-1	Freedom	
1-2-3	Expansion	

Partition focuses on the essential difference between higher and lower worlds which are *becoming more existential*. The domain of harmony describes what can happen in a lower world due to a higher world: the arising of journeys, myths, gods and demons, and, indeed, angels. In the Bible or the Qu'ran, angels are agents of the Universal Will in contact with the human world. The narratives that result translate this world into our outward world of (1) separations, (2) journeys, (3) struggles, (4) dramas, (5) resolutions, and (6) reunions; these are all another way to see World 6 in our own terms.

Bennett also had insights into the early Bible.

The creation myth of Genesis is really an account of the creation of life, not of the earth, and it presents a picture of a transition from a state of chaos to a progressive ordering that culminates with a sexual being. The transition is ascribed to the action of the **Elohim**, the powers of God, not to God, or **Jehovah**. . . .

The main creative work is one of **separation** or **partition** [where] the light is separated from the darkness, the firmament from the earth . . . making possible an action between the separated parts that is fruitful, because out of their blending comes something they did not have before their separation that comes from the creative act. Genesis speaks in the language of its time of the principle of the law of threefoldness: the higher blends with the lower to actualize the middle.[6] (Emphasis in original.)

This may seem far from the buildings of previous chapters, which were models of the Earth's creation within our time-world, but this partition of will is the precursor to a world in which there can be separation of places, objects, and beings, and in both space and time these will be intelligible as an elaboration of patterns in Eternity, preexisting before their creation.

The writers of the Bible could not know our discovered model of galaxies full of stars and of a universe full of galaxies. Their idea of the Creation was, quite rightly, from the Sun, which is a *secondary* cosmos. Bennett notes:

> Between the sun and earth (or planets) is the **Elohim**, the **Celestial Hierarchy** or **Heavenly Host**. Gurdjieff used the latter term which figured very strongly in the early Christian liturgies* as **All the Company of Heaven** . . . the Elohim in Genesis representing the working of the third force.[7] (Emphasis in original.)

A partition between worlds removes something from the higher world by duplicating its triads, with one of its essential forces being "existential," or impure. If one looks at the triad of involution 1–2–3, Bennett relates this to the Self as having a partner 1–2–3,*[†] the star denoting the replacement of the essential force of reconciliation by an existential one in World 12, leading to its identification with the Sun. In table 9.2, the Universal Being is purely essential (1–2–3) but, when the last force (3, or reconciling) is rendered existential, the True Self (or Individuality) belongs to World 12, a world with six purely essential triads and six altered in just one force—the last force of each triad permutation (see systematics.org s.v. Triad: *reconciliation is the impulse that enables the dynamism "to be."*)

The True Self is Cosmic Individuality, whose third force of reconciliation acts upon existence. To generate World 24, the two triads of the True Self are both duplicated to have their second forces made existential (fig. 9.6), making

---

*Bennett's footnote says, "The liturgy was a very remarkable construction for which a group in Cappadocia was largely responsible. It was probably put together in the third century and in its structure of three parts probably derived from the old Zoroastrian liturgy. Gurdjieff claimed that it was one of the most extraordinary **legominisms** in the world and all the secrets of the universe are contained in it." Bennett, *Creation,* 104. (Emphasis in original.)

†The star symbol (*) is part of Bennett's terminology, indicating here that force 3 (the reconciling force) has become an existential, not an essential force.

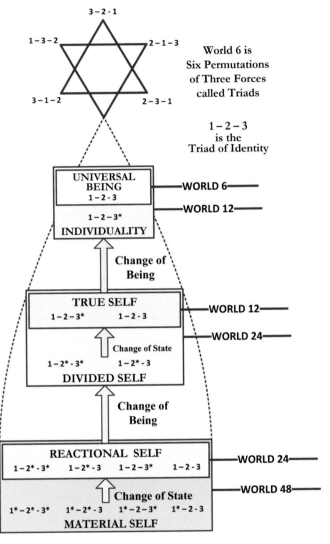

World 6 is
**Six Permutations
of Three Forces
called Triads**

1 – 2 – 3
is the
Triad of Identity

FIGURE 9.6. The place of Selves in Worlds, illustrated by the law of involution 1–2–3.

Adapted from J.G. Bennett, Deeper Man, *fig. 8.2.*

twenty-four laws in all, and these are (in ourselves) the domain of the Divided Self. This is, to recall Bennett's quote, "the action, which sets a process in train by the relatedness of impulses."[8]

World 48 repeats this duplication, this time with the first force made existential, creating the psychological worlds of the Reactional and Material Selves. The pure versions of each triad are still available in the lower worlds, but their significance is rarely grasped and is so lost to ordinary human experience.

In the creation of the macrocosmic worlds, the double duplication of triads creates a dyad between (in this case) the Sun (as Father, World 12) and the Earth (as Mother, World 48). This dyad must be mediated by another pair of terms that are (in this case) the biosphere (World 24) as the chosen active principle of living things (relatively independent systems that are autonomic in World 24) and the Heavenly Host as the Sun's planetary world as a cosmic system (the hypernomic Creation of World 24), the biosphere being a *lateral octave* for the Gurdjieff of *In Search,* starting in the sun.

Though not directly involved in Life, the planets are defining time on Earth as a framework condition through their chosen vehicle, the Moon, as is seen in the harmonic model/matrix of the inner and outer planets. This matrix is limited by the large number of YHWH, 777,600,000, or 60 to the power of 5.*

So why this large number?

> The Sun is not God, but it is a creator, particularly the creator of life and the Earth, the Mother of Life, who bore life in her womb, gave birth to life and nourishes and cherishes life. When it comes to the Heavenly Host, we cannot see them as we see the Sun and the planets. The planets influence the formation of Her children [just as] the human environment influences the child. Life receives an important part of its forms and possibilities from the other planets of the solar system.
>
> The planets [can be called] "abodes of the Heavenly Host." Yet [also] the Earth, "life's prison." Between these two conceptions is Earth, "the bearer of life." The Heavenly Host appears as the enabling factor which integrates life and the planets [while] Earth as Life's Prison is the manifestation of the denying principle. It is through **this** that transformation and return to the source is possible. Earth as Life's Mother simply passes on the creative process through life and does not provide the way of return.[9] (Emphasis in original.)

This cosmological principle of partition is complementary to the numerical and geometrical models required to manifest levels of Being. The form of the cosmos,

---

*This is the number of Greek geographical inches (1.01376 inches) between the North and South Poles of the mean or spiritual Earth. In the ancient Greek world, Apollo had this same number of 7776, or 6 to the power of 5, a "head number," since the zeros in its tail can be dropped for brevity, as in Plato's *Ion,* with "Apollo in it," which is composed of 7776 syllables.

especially at the detailed level at which humans actually live, shares forms inherited from all the higher worlds. In this sense, they are repetitions of them into a new, more specific setting.* Similarly, the content of life on Earth is time- and space-factored, that is, matched in its content and form by the world of the Heavenly Host, where "angelic transmission" has made the heaven and Earth seen by megalithic astronomers through compatible models of the higher worlds using "sacred" numbers that are factoring the time cycles and spatial geometries.

A higher world is of necessity simpler than a lower world, hence a higher world is freer because it embodies half as many laws. So in spiritual and mystic texts, there are narratives involving gods just as if these occurred on Earth because the forms of situations in both worlds resemble each other, just as the dream world has elements found in waking life.

Five and the Essence Classes

One hundred years ago, in his "Diagram of Everything Living" (henceforth Diagram, see fig. 9.7), Gurdjieff presented (in 1917) his ideas about beings and their transformation of human and cosmic energies. Ouspensky tells us that Gurdjieff said this diagram was:

> "still another system of classification . . . in an altogether different ratio of octaves . . . [that] leads us beyond the limits of what we call 'living beings' both higher [and lower] than living beings. It deals not with individuals but with classes in a very wide sense.
>
> "Each square denotes a level of being," he said. "The 'hydrogen' in the lower circle shows what the given class of creatures feeds on. The 'hydrogen' in the upper circle shows the class which feeds on these features. And the 'hydrogen' in the middle circle is the average 'hydrogen' of this class showing what these creatures are.
>
> "The place of man is the seventh square from the bottom or the fifth square from the top. According to this diagram man is 'hydrogen' 24, he feeds on 'hydrogen' 96, and is himself food for 'hydrogen' 6. The square next below man will be 'vertebrates'; the next 'invertebrates.' Invertebrates are 'hydrogen' 96. Consequently man feeds on 'invertebrates.'"[10]

---

*Called "similarities-to-the-already-arisen" by Gurdjieff. *Tales,* 758.

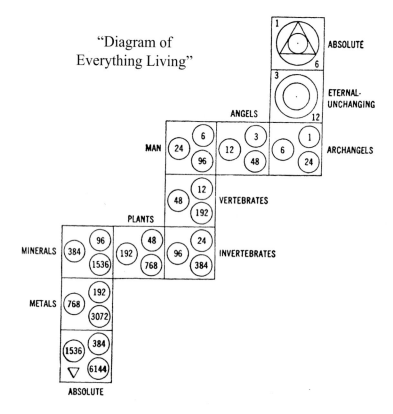

FIGURE 9.7. The "Diagram of Everything Living."
Presented by Gurdjieff soon after February 1917.

There is a pattern in the numbers of the diagram of figure 9.7. Ouspensky never forgot it, and by the 1950s, Bennett, who was Ouspensky's pupil in the 1930s, could explain the limited population of essence classes shown as populating the universe. These, expressed using the numbers of the Worlds, had a fivefold constitution within the world of will. Each class belonged to a World number that defined its unique role in the universe of will that is centrally placed to four adjacent worlds, all of which are then terms that are different but acting on the same level. The will of an essence class is always defined by the number 5, which can be called its pentad,[11] offering its specific world the "significance" or "potential" within that world. This is the pattern for each type of system within Bennett's Systematics: each has (a) a unique number and (b) a systemic attribute. For example, the triad's 3 terms express "dynamism."

In Gurdjieff's Diagram, the lower nine square units all have three circles filled

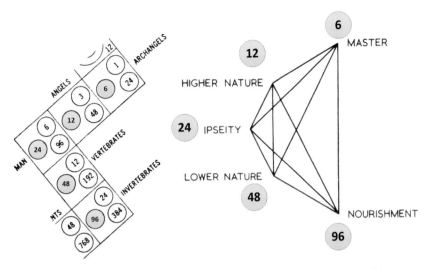

FIGURE 9.8. The five-term Systematic (pentad) as conceived by J. G. Bennett (*right*), and the section of Gurdjieff's *Diagram of Everything Living* relating to the human essence class (*left*). The yellow "hydrogens" naturally correlate the human to the beings above and below.

with three world numbers. These define levels of hydrogen corresponding with (*center*) the being itself, (*bottom*) what it needs to eat, and (*top*) what is nourished by it, noting that larger numbers are beings of lesser worlds. One can immediately correlate that the two right-hand elements give the **outer significance** or context of the being. For the class of human, World 24 must be doubled twice to reach the humans' food of 96, and the intermediate hydrogen is that of the class below those humans: the vertebrates. Similarly, what those humans nourish is doubly halved as 6, and 12 is the level of the next higher being to the human essence class. Thus, the **inner range of potentiality** of the essence class man is between the animal (48) and the angelic (12). Bennett's pentad was clearly prefigured in the Diagram, evolving then into Bennett's system of twelve essence classes. Bennett's excellent "The Great Laws," appendix II of Bennett (1973), should be read alongside it.[12] The group receiving the Diagram might have equated angels and archangels to the planets and stars of the Ray of Creation, so Gurdjieff qualified it: "This diagram will not be very comprehensible to you at first. But gradually you will learn to make it out. Only for a long time you will have to take it separately from all the rest." For while essence classes were defined by their structuring of Will, they manifest as Being, and their Function is to eat and be eaten.

The Pentadic Structure of Cities

Bearing this in mind, it is interesting to put flesh on the bones of the pentad by looking at the nature of the ancient Middle Eastern cities, with an emphasis on the city-states begun by the Sumerians in Mesopotamia (fig. 9.9). Cities had a god that was a king (archangels), while their existence depended on the **surpluses** created by plowing the soil (as invertebrates do) and farming with animals (the vertebrates), and these two techniques enabled more than subsistence farming. Cities, and the farming needed for surpluses, naturally created property ownership and trade, especially as a social hierarchy developed of **leaders and specialists** (angels). The leaders and specialists came to depend on information in order to preserve know-how, ownership, and trading accounts, while heroic myths and historic legends reinforced legitimacy and established behavioral values. The land and the city became fused in a sacralization of the landscape, connecting the earth to the heavens and gods. Farming methods meant that, in some fashion, the solar calendar needed to run alongside the prehistoric lunar calendar. War between cities became conflated with sacrifices to the city gods.

The discipline of the five-term pentad gives insights as to what the essence class of a city is. Cities must be a development of the human essence class

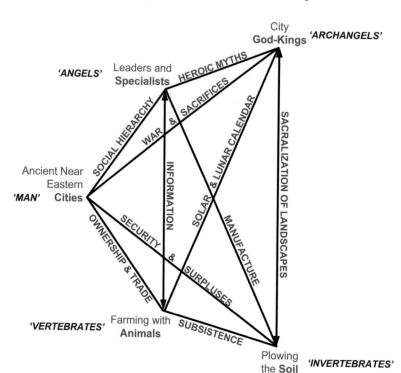

Figure 9.9. Application of the pentad to the reality of ancient Middle Eastern cities.

since they facilitate some of the potentials (the inner range) of the human essence class through creating proxies for the angels and archangels of the Diagram we have been considering. This would also apply to Greek cities and the Mexican cities of the Olmec and Maya. By creating proxies for the higher essence classes, useful specialists came into existence, such as tradesmen, soldiers, supervisors, and scribes. These "upper classes" represent the higher worlds, most clearly when religious ideas and specialisms arose within them. Their work concerns the heroic myths, the sacrifices and sacralization of time, the calendar, space, and the landscape—all of which emanate from the god-king. Bennett called this period in the ancient world the Hemitheandric epoch (3200 to 800 BCE), when special humans—heroes and priest-kings—were believed to have superhuman attributes and powers enabling them to act as intermediaries between the gods and ordinary people.[13]

NINEFOLDNESS AS A UNIVERSAL SCHEME

In *Tales,* the law of ninefoldness was God's primary innovation in combining his own "system of functioning" with the law of seven and the law of three.[14] Formerly, they worked independently of one another. The new threefold law rejected the linear simplicity of one-thing-leads-to-another and introduced a triangular threefold order (fig. 9.10).

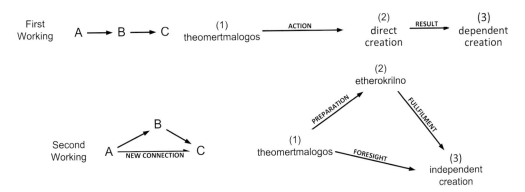

FIGURE 9.10. Bennett's graphic shows God's change to the law of three. Simply put, theomertmalogos is the "Word God" and etherokrilno is the creative number field.

Adapted from Bennett, Talks on Beelzebub's Tales, *1977, 51–57; reprinted 2007, 62–68.*

In this threefold law, the Universal Will (1) must create something out of "nothing," (2) with the aim to achieve a state of affairs in the future, (3) an *independent* creation. This pattern came to Gurdjieff while he was fighting for his life after a bullet wound, during which he asked why God appears to sacrifice humanity to apparent helplessness within a mechanical world. It then occurred to him that (a) God sacrificed something of value to achieve an independent creation, but that (b) he himself must also achieve this new type of working to become independent of the mechanical world. In *Tales,* he uses God's dilemma and its solution to explain how the microcosm has the same dilemma, thereby solving the enigma of what the microcosm, the image and likeness of God, was for.

> Our Maker Himself, in the name of all that He had created, was compelled to place one of His beloved sons in such an, in the objective sense, invidious situation. Therefore I also have now for my small inner world to create out of myself, from some factor beloved by me, an alike unending source.[15]

This change in the law of three explains why there is no supreme god of direct affirmation in our world, contrary to the assumptions of the faithful. God is active only in the reconciling force, within the relationships of the independent creation. This is especially true in the new form of the law of sevenfoldness, in which "HE actualized the greater change in the law of the sacred Heptaparaparshinokh"[16] formerly of seven but now ninefold. In Bennett's Systematics, the systemic attribute of ninefoldness is **harmonization**: "The structure represented in the octad is still lacking in concreteness, for it does not take account of hazard and uncertainty.... The ennead is the first in the progression of systems that can take account of both purpose and the uncertainty of its fulfillment."[17]

The natural arising of hazard (that is, the uncertainty generated within existence in time and space) prefers accidental results, as with the entropy that runs down, rather than improves, the established order. The mechanical World 48 is bracketed by the law of accident below (World 96) and the angelic above (World 24), the latter world being also the higher nature of the human essence class in Gurdjieff's Diagram.

The new law of three became part of the new law of seven (whose systemic attribute is transformation) in that (a) the denying force could improve the quality of receptivity between an octave process and external processes, while (b) the reconciling force should require the participation of consciousness for processes to achieve their aim. Only then would processes become transformative and able to overcome uncertainty, or they would fail and need to start again. This law becomes the means whereby the human essence class must evolve through being a proxy for God's will within the creation.

The first force, of affirmation, remained common as the Universal Will, and so the law of sevenfoldness shared oneness with the law of threefoldness so as to produce a composite system of ninefoldness, not of ten. The form of this law is both the Enneagram diagram presented in *In Search* and the Heptaparaparshinokh described in *Tales*. The same system was not evolved between them, and the descriptions are mutually supportive yet outwardly different.

THE ENNEAGRAM OF LIFE

The Enneagram was presented as a universal principle and was thus relevant to the working of the microcosm. It can therefore be compared to the geometrical Equal Perimeter Model, where some church pavements expressed the human microcosm within the whole design as microcosm, a reflex of macrocosm, the universe, and conscious being. Ninefoldness may be a formula for the life of conscious individuals and the reciprocal maintenance of all living processes, but the nonliving material processes obviously do not have to adapt. This means that ninefoldness is a weak property in natural selection but a strong potential for human transformation—an enneagram for Life on Earth.

The diagram's musical progression must not be confused with regular music. The application of musical notation such as *do-re-me-fa-sol-la-si-do* implies an ascending octave within the universe, perhaps between the Worlds. *In Search* presented the Ray of Creation and other similar octave diagrams labeled according to increasing scale. However, the harmonic model of the Moon and the planets is a vision taken from the harmonics of Being and of Eternity whereas the Enneagram is a structure of will.

The numerical limits of the Worlds {3 6 12 24 48 96} are interesting when seen in their true context, that of the cardinal set of the number 3 doubled up to 96. Their growth in number allows musical tones to arise. For example, World 12, as an octave limit, can support Plato's demiurgic model of (6:8::9:12), and we know that Life was said to be a side octave, emanating from the Sun, then identified as World 12. Suppose then that a pattern in Eternity, expressing what Life is, were to extend between worlds 12 and 96, as in figure 9.11.

It can be no coincidence that the major scale emerges between the planetary world number of 24 and the Earth of 48, then is repeated to the Moon's world number of 96 in the cosmology of Gurdjieff's Ray of Creation. Figure 9.11 teases out the musical interval ratios between the early numbers up to 96, making the emergence of these scales visually obvious. This implies a design methodology based on the minimum limiting number necessary, something also seen in

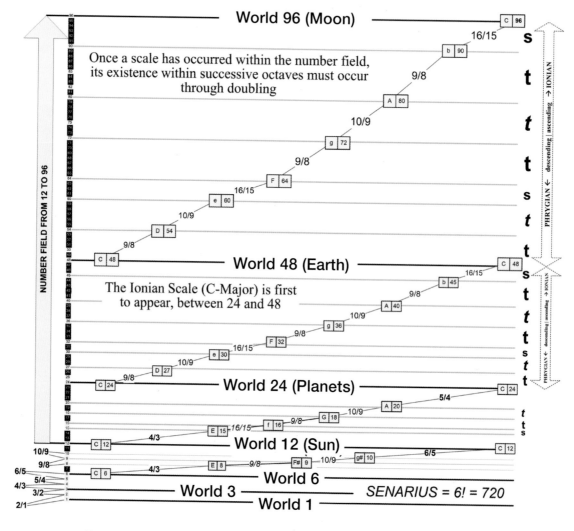

FIGURE 9.11. Early numerical emergence of the major scale between the planets and the Moon. On the right, *s* stands for semitone and *t* stands for tone.

the harmonic model for limit 720. That the lowest possible limits have been used confirms their use by the Universal Will, a requirement achieved by the Moon to extend but also limit the extent of the Creation with a large planetary Moon, when there is one. Otherwise, that limiting role is played by the center of a planet in a scheme called Polormedekhtian rather than Keschapmartnian.[18] (The Polormedekhtian planets have no large moon while the Keschapmartnian have a large moon.)

Instead of studying these numbers upward, they can be drawn as descending, in the sense of an increasing depth to the Creation. In that case, one can see that the Enneagram, shown alongside the range from 12 to 96, shows the Worlds as corresponding to its points (fig. 9.12).

The bottom of the equilateral triangle lands in the third and sixth stopinders,* suggesting that it is Life that will eat and be eaten *in order to be*

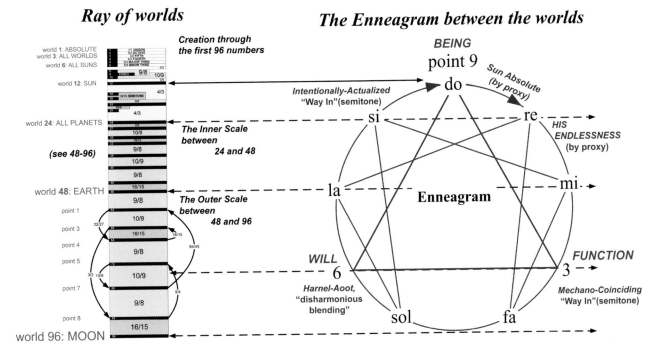

FIGURE 9.12. This diagram shows the correspondence between the range of numbers from 12 to 96 and the levels/points of the Enneagram. The terminology of *Tales* is also shown.

---

*Stopinders were defined by Gurdjieff as "'gravity-centers' of the fundamental 'common-cosmic sacred Heptaparaparshinokh.'" This and other terms can be referenced using the glossary at gurdjieff.org.

autonomic. Gurdjieff's Trogoautoegocrat, the law of reciprocal maintenance, could have been a third way of looking to those stopinders, providing that the Enneagram and Heptaparaparshinokh, which extend between these two world limits, 12 and 96, exist as a side octave of "there and back again."

- World 12 is on the level of *do,*
- World number 24 is on the level of *si* and *re* (the notes adjacent to *do*),
- World 48 is aligned to *la* and *mi,* another step away from *do,* while
- *sol* and *fa* are down with the Moon as World 96, the tritone.

This horizontal layering is similar to the compositional rings used to structure ancient texts, a technique called ring composition.[19] It would indicate that the worlds are symmetrically defining levels of concrete development and that the Moon is at the middle and bottom, where the central meaning of a story, here the limit, is reached. This location, called the Turn, is where the descent into necessary structures, within a narrative, turns back toward the higher worlds on what is called the path of return. This central point was traditionally identified with the god of fire, who burns up illusions and brings wisdom. It is also the phoenix that is reborn from the ashes.

Between 48 and 96 lies a descending major scale that repeats the same scale between 24 and 48. Points 6 and 3 land on its third and fifth stopinders, implying that reciprocal maintenance is related to these points, from the law of three now additional to the law of seven. They may be the "two birds" of the Rig Veda and the Upanishad,* one eating the tree's fruit and the other looking on. Something must be given if something else to going to be received. The diagram could also explain the dictum of Heraclitus (died c. 473 BCE), that "the way up and the way down are one and the same." The reversal of forces through an act of will is associated with point 6. The Moon, at this midpoint between points 3 and 6, is the counterweight for the creational Worlds 12, 24, and 48, forming a $\sqrt{2}$ tritone to the *do* at World 12, to limit the development of the intelligible universe as does Saturn in the World Soul. As Bennett obliquely put this in the following quote:

---

*"The Tree of Jiva and Atman appears in the Vedic scriptures concerning the soul. The Rig Veda samhita 1.164.20–22, Mundaka Upanishad 3.1.1–2, and Svetasvatara Upanishad 4.6–7, speak of two birds, one perched on the branch of the tree, which signifies the body, and eating its fruit, the other merely watching." Wikipedia, "Tree of Jiva and Atman."

Earth as Life's prison is the manifestation of the denying principle. It is through **this** that transformation and return to the source is possible. Earth as Life's Mother simply passes on the creative process through life and does not provide the way of return.[20] (Emphasis in original.)

It may well be that Gurdjieff was preparing a future path for humanity relating to the higher worlds through understanding their numerical structure, just as the traditional models were becoming swamped by modern functional approaches to the existing world—a seemingly irrepressible materialism. All three aspects of Function, Being, and Will are always present, but two can be quite hidden due to the "blindness" of the modern psyche to the mysterious or subconscious dimensions that are Bennett's Eternity and Hyparxis.

SYSTEMATICS IN THE SKY

The language of angels, found in many ancient monuments, represented a great Being in the sky governed by numbers of days, months, and years. This remarkable view, achieved by a now ancient people, is not given us by our own sciences and technologies that, through attention to physical processes, have achieved a language of Function, finding the world not held together by numbers or geometries but by self-organizing interactions between forces, objects, and substances. These two views then, of Being and then of Function, have taken us even further from understanding the initiating will behind the Earth's tight relatedness to the sky and its time periods. It seems unlikely that the view of great Being can return unless it suits a universe that appears detached from the human world except in being part of the world that created us. But, being part of the whole destined to understand the world, each individual must realize the act of will that understanding is.

J. G. Bennett identified that one must agree to what is going to be understood.* This poses the problem of how to agree if not already understanding. In order to approach understanding, Bennett developed a language of will

---

*Like the word *consciousness, understanding* has a different meaning within Gurdjieff's ideas than in ordinary life where understanding means comprehending or even "familiarity with" something. In this case, we can turn to one of P. D. Ouspensky's great aphorisms "To understand is to agree" to see that understanding is beyond disagreement: you either get it or you don't.

called Systematics—encountered in the last chapter in the discussion of the human essence class and the ancient Near Eastern city that could be understood using the pentad, a fivefold structure of will capable of understanding the individual identity or selfhood of something that exists in its own right. Being a language of will, this can be used to see that the time-world of our sky is a structure of will whose prime mover is the Moon.

There are two primary structures expressed by the Moon. Its simple outer manifestation is its phases of illumination by the Sun during the lunar month. Behind this lies the inner manifestation resulting from the retrograde movements of the lunar orbit's nodes, touching the Sun's annual path in two diametric places called ecliptic nodes. The latter long cycle, seen from Earth, causes the Moon to ride higher than the Sun in the sky for a period of 9.3 years, and below for the same time, just as the sunrise is north or south of east, for one half a year. Two pentads can be formed for each of these manifestations, revealing that celestial time is all about something perpetually catching up with something else.

The significance of the lunar nodes is shown strongly but only occasionally by eclipses of both the normal presence of the Sun or Moon. Many monuments from many periods and regions of the Earth seem to reflect the cycles caused by the lunar nodes, within geometries of triangles (and the rectangles that can contain them) and the Equal Area Model, within a language of Being that cannot say how or why such time patterns came to be. The pentad of the nodal periods can approach this causal world (of will) where things had a purpose (fig. 9.13*A*).

The cycles of the Moon's illumination can also be compared (fig. 9.13*B*) showing very different numbers while the same form is evident; the pattern may be imposed by the pentad but recognized as uncanny, as if the essence class of celestial cyclicity has been revealed. Notice also that the numbers all add up arithmetically within a structure, despite the different paths between the time cycles within each pentad.

On one level everything is bound to add up, but on another level each system generates a ratio between the thing itself (the lunar year or eclipse year) and the solar year, as seen in both diagrams. We have encountered all these numerical realities within sacred geometry, but the pentad gives a different view due to its language of will. The form of the pentad may be the Illumination Bennett saw as coming from the Sun to the angelic host, where the higher intelligences

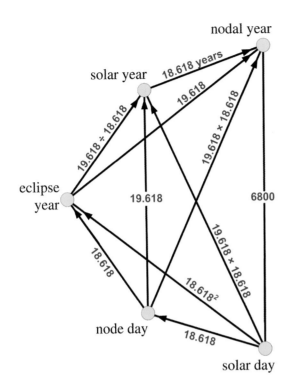

nodal year

solar year — 18.618 years →

19.618

19.618 ÷ 18.618

19.618 × 18.618

eclipse year

19.618

6800

18.618

19.618 × 18.618

18.618²

node day

18.618

solar day

FIGURE 9.13A. The pentad of the nodal cycle.

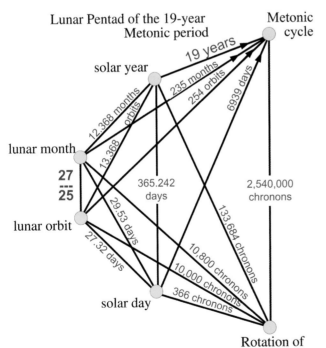

Lunar Pentad of the 19-year
Metonic period

Metonic cycle

solar year

— 19 years →

12.368 months

13.368 orbits

235 months

254 orbits

6939 days

lunar month

$\dfrac{27}{25}$

365.242 days

2,540,000 chronons

lunar orbit

29.53 days

133.684 chronons

27.32 days

10.800 chronons

10.000 chronons

solar day

366 chronons

Rotation of
Earth
(CHRONON)

FIGURE 9.13B. The lunar pentad of the Metonic period.

associated with the planetary system were free to play with "the train set" and where the pentad would be like their simple control panel.

The simplicity offered by so few as five terms seems purely coincidental to any modern awareness of these terms, whose names are somewhat familiar. Even Kepler could not find this though he intuited something like it by, again, using geometry. It is only when the language of will is used that one can see the form of time as being its purpose: of creating a coalescent system in which the eclipse, lunar, and solar years are held in a vice-like grip of a gravitationally recurrent N-body system (as modern science would call it). Here one sees a "war against time" in the sense that order is being preserved against the chaos of accidental events, that is, events outside the structure of the Universal Will. While accidents still appear to happen within the world of time, there may be nothing accidental about them when seen from a higher world.*

The modern perspective of functional awareness, because of its nonacceptance of what would be seen, makes it unlikely that human beings can understand something so fundamental, so simply. Only will can understand, yet it remains elusive or fleeting without a rigorous discipline like Systematics, which may hold the key to the transformation of the microcosm itself. The fantastic time world, enabling life and the microcosm to evolve, does not provide or guarantee humanity's further development, since that development requires human acts of will, and these are rare unless sought. The pentad of the human essence class shows the angelic as our higher nature and beyond that the potential to go beyond the Universal Will. That is, only human acts of will can transform a human being and achieve that goal.

We have identified sacred geometry as a language of Being, which first arose through study and counting of sky cycles in their duration and then arrived at spatial geometrical forms. From this arose the sense of being inside a great Being or its creation. Modern science and its attendant technologies now speak a different language of Function in which everything in nature is reduced to the operation of physical laws and the acquisition of great knowledge of, and significant power over, the physical universe. It was the deep past that gave us the minds to build modernity, but the realm of the microcosm, since Pythagoras

---

*"In real art there is nothing accidental. It is mathematics. Everything in it can be calculated, everything can be known beforehand." George I. Gurdjieff in Ouspenky, 1950, 26–27, noting that the time world, as we have seen, is a work of objective art.

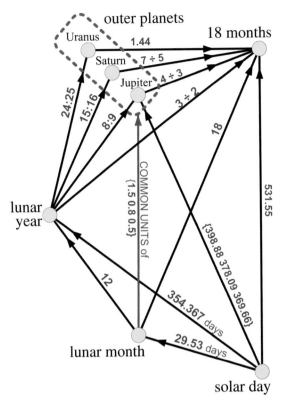

FIGURE 9.14. The triple pentad of the harmonic model of outer planets in harmony with the Moon.

first glimpsed its existence within numbers and sacred geometry, points toward a destiny involving the language of will and its ability to understand ourselves and our world in a perpetually new way.

The pentads of figure 9.13 show some part of the will that created time on Earth, and of what the planetary world was made to do as a whole device. Holistic thinking is much vaunted but is hard to communicate without a suitable language of will that lies in our future rather than the present or the past. For it is will, universal or individual, that creates the future and effectively is the future. But will can only be known by its manifestations in the present and who we become because of it.

Emergence

GURDJIEFF'S VISION OF MANY WORLDS, each created out of a higher and simpler world, provides a reason why the Earth holds uncanny numerical coincidences and why ancient builders sought to reproduce these then sacred measures of the Earth and Sky in buildings. The connectivity between these worlds ensures that the Creation is held together by numerical relationships that act as a glue and as conduits for the exchange of energies between the worlds of a single whole.

The Universal Will, which brought about the universe, sought to limit its development through the creation of life and, in particular, of each human life that holds the potential for Cosmic Individuality. This became the vision of the people of the Abrahamic religions, who evidently inherited the knowledge that the planets were harmonized to the Earth through the lunar year, a relationship that expressed the oneness of God through the musical octave. The Bible, written in 600 BCE, records four major influences from the ancient Middle East: the Mesopotamian, the Egyptian, and the Persian civilizations, and the matriarchal Mediterranean religion, and it blended many religious ideas into one that rejected many gods and recognized only one.

Around the same time Pythagoras realized, through his travels through those same areas, the primacy of the octave, the music of the spheres, and the significance of numbers within cosmology. The increasing individualism of the Classical Greek world, leading to an early form of democracy, reinforced the idea of a singular God as savior—an Indo-European idea congruent with a singular act of Creation leading to the universe. The coming of the Son of God *made concrete,* through incarnation, the biblical idea that humans were the children of God, resulting in Christianity.

The making of buildings that reflected the higher worlds was already a widespread tradition of giving a part of the Earth the intelligible form of the planet itself, or of Earth's geocentric patterns of time. The doctrine of the microcosm vis-à-vis the macrocosm (after Pythagoras) revealed the human being *as being like a sacred building,* already reflecting the cosmos while also reflecting God in the sense of completing the universe. In this way, both sacred buildings and the human body were modeled on a sacred universe through the connective power of numbers. This would allow the Kaaba to represent *the whole of mankind* held within the building that God had made and that was connected to the higher worlds, just as the human was a similar building, but then based on selfhood within the universal scheme. But these buildings were not automatically seen to be what they are since, "Except a man be born again, he cannot see the kingdom of God."[1] Modern science lacks this vision despite its amazing power to interrogate the universe.

To all that went before, the simplest solution is that the human mind is a reflection of the demiurgic mind of creation and that the demiurge has become human. We need minds, but the modern collective mind is identified with the endless extension of the material world, led by desires we do not really own but we absorb in childhood and general life. Juxtaposed is the impulse to be ourselves, but then to what purpose? The numerical view of our place in the universe implies that there is an emergent role for individuality that is cosmic and not collective. This numerical view has developed in the last century alongside ideas about self-development, and these two themes are compatible and necessary for the understanding of ourselves as crucial to the operation of the universe itself as *the chosen mechanism* for there to be heaven on Earth.

A Developmental History of Metrology

METROLOGY HAS APPEARED IN MODERN TIMES (phase five below) in reverse order, since humankind saw the recent appearance of many measures in different countries as indicative that past cultures made up units of measure as and when they needed them, perhaps based upon lengths found in the human body. But this soon breaks down under scrutiny because the measures called after different regions all have systematic ratios between them, such as 24/25 feet (which as a foot is the Roman) and 6/5 feet (which is an aggregate unit, a remen), and the size of humans is quite various between regions and within populations. As stated in the main body of this book, the notion of measures from different regions was called historical metrology. This framework began to break down when answers appeared as to why the different regional feet were related, not only to the English foot as equaling one for each ratio, but also to the fact that the units of measure were often seen to divide into the size and shape of the Earth (leading to our phase four)—then called ancient metrology.

Another aspect of measures was their ability to approximate important, otherwise irrational, constants (our phase 3), such as π, √2 and even *e* in the form of megalithic yards, which are close to 2.71828 feet, the numerical value of *e*—the exponential constant. The earliest megalithic yard was almost exactly that number of feet—derived from an astronomical count over three lunar and solar years in day-inches (chapter 1) leaving a 32.625-inch difference between these years (our phase one); those 32.625 inches equal 2.71875 (87/32) feet.

The gap between the first and second phases of metrology seems to be the gap in time between the megalithic in Brittany and in Britain. Only as the metrological purpose of more megalithic monuments becomes clear might one be able to know more accurately, but British metrology, in choosing a megalithic yard of 2.72, was able to factor the nodal prime number of 17 within its counting. At Le Ménec's western cromlech, Brittany could use a radius of 17 megalithic rods (6.8 feet) to have a count of 3400 megalithic inches across a diameter. Later, Britain could use 12 such rods to model the lunar year of 12 months and then also count 15 rods as 3400 shu.si, a small digit known to historical metrology as dividing the 1.8 foot (the double Assyrian foot of 0.9 feet) into 60 parts. The shu.si (0.03 feet) divides into many foot modules (see p. 112), and notably the English yard as 100 shu.si so that 68 yards equals 6800 shu.si. This length, 204 feet, enabled the nodal period to be counted by the clava culture at Balnuaran in Scotland.

There is a particular need to regularize this subject through the gathering of more examples of metrology's past applications. One must recognize that those responsible for our present knowledge of it have largely passed away, and those in academia are not going to rewrite history in order to impartially reassess whether their own approach to ignoring it can still be justified, especially when they are not preserving the metrology within monuments because they can't see it as a signal from the past.

Overview of Megalithic Units of Measure

At least five specific MYs have emerged from the counting applications within megalithic monuments:

1. The proto-megalithic yard (PMY) of 32.625 day-inches, emanating from an original day-inch count over 3 solar and 3 lunar years, as the difference in their duration (chapter 1). This belongs to the world of inch counting.

2. The Crucuno megalithic yard (CMY) of 2.7 feet: We saw that, by the factorization of 32 lunar months as 945 days long, the lunar month (as 29.53125 days long) can be represented by 10 MYs of 2.7 feet, where the days in such a count are the Iberian foot of 32/35 feet. This I call the Crucuno megalithic yard, though, in the historical period, this foot came to be called the root foot (27/25 feet) of the Drusian module,

which, times 25, is then 27 feet. The astronomical megalithic yard AMY (*next*) is 176/175 of the CMY.

3. The astronomical megalithic yard (AMY): In Britain, this is 2.715 feet (32.585 inches) long, giving $N = 32.585$ for the actual $N:N + 1$ differential ratio between the solar and lunar years. When representing lunar months over a single year, the excess becomes the English foot of 12 inches—the megalithic, now English, foot. From this one sees that every AMY on the base of the Lunation Triangle defines an AMY plus 1 inch on the hypotenuse above it (length $N + 1 = 33.585$ inches), as the duration 1 mean solar month. The AMY can appear as an integer when the CMY defines a radius because it is 176/175 of the CMY.

4. The nodal megalithic yard (NMY): Used in Britain. Thom's *Megalithic Sites in Britain* gave the megalithic yard as having had the value of 2.72 feet as "the" MY, based on integer geometries within stone circles and some statistical methods applied to other inter-stone distances Thom had measured. Its value evidently derives from its relationship to the nodal period of 6800 day-feet because $2.72 = 6800/2500$, where 2500 feet is half a metrological mile of 5000 feet. For this reason, I now call it the nodal megalithic yard (NMY), which contains the key prime number 17 in its formula 272/100, 272 being 16 times 17. Its megalithic rod (NMY times 2.5) of 6.8 feet factorized the nodal period of 6800 days: 15 rods gave 102 feet (3400 shu.si) and 30 rods gave 204 feet (6800 shu.si), the shu.si being $204/6800 = 3/100$ feet. It therefore appears that the NMY, its rod of 6.8 feet, and the shu.si had a raison d'être in the British megalithic period that was focused on the later problem in astronomy of counting the days of the nodal period.

5. The later* megalithic yard (LMY): Seen at Stonehenge and Avebury. Thom in 1978 published a *new estimate* for the MY as 2.722 feet. Unbeknownst to Thom but lurking within his own error bars was a further development of the AMY which, times 441/440, would locate his value within ancient metrology as 2.716 feet, 126/125 of the CMY. The CMY is clearly the root value (in Neal's terminology 2.5 root

---

*in the context of Thom's work.

Drusian of 27/25 feet) and the AMY the root canonical value, while this LMY is the standard canonical value.

All of these different megalithic yards had their place in the megalithic people's pursuit of their astronomical knowledge. Noting the role of the shu.si in compressing the length of a nodal count to a mere 204 feet, Thom's NMY of 2.72 is the key to how its length of 3/100 feet was arrived at. The shu.si of 0.03 feet (0.36 inches) surprisingly divides into many of the historical modules of foot-based metrology.

Historical Module	Foot Ratio	In Shu.si	Notes
Assyrian	9/10	30	Carrying the sexagesimal (base-60) system of the Sumerians.
Roman	24/25	32	
Inverse Byzantine	99/100	33	Times 3 gives 99, a yard minus one shu.si.
English	1	33.<u>333</u>	Times 3 gives 100 shu.si in a yard.
	51/50	34	Divides into the nodal period. The difference between 80 and 81.6 feet and between 90 and 91.8 feet at Seascale, where 91.8 locates the Jupiter synodic period.
Persian	21/20	35	Its remen (6/5) is 42 shu.si.
Drusian	27/25	36	The CMY is root of the AMY and the LMY.
Remen	6/5	40	Half-remen of 20 shu.si as ideal form of the equal perimeter.

FIVE PHASES FOR METROLOGY

Metrology as a single system was based on the number 1, which was then realized as the English foot (Robin Heath, 1998), which then became related to all the foot modules of the ancient world—through a range of simple fractions.

There were, therefore, phases in the evolution of ancient and then historical metrology. I can see five right away.

Phase One: An Inch-Based Metrology for Astronomical Counting[*]

Primordial measures arising from the conduct of astronomy in the megalithic period included the English inch used to count days at Le Manio, Carnac; the Carnac megalithic yard of 261/8 inches arising from Le Manio's three-year count, forming the Lunation Triangle; and the English foot arising from counting the Lunation Triangle over a single solar year as lunar months using the CMY per month.

To form the English foot required definite steps that were necessarily taken through megalithic astronomy and findable in the monumental record as (a) the use of the inch to count days over 3 solar years,[†] (b) the use of the differential length over 3 years to count lunar months rather than days, and (c) the counting over a single year to find an excess length of the English foot, which still has 12 inches because the lunar year has 12 months.

Phase Two: A Foot-Based Metrology for Astronomical Counting[‡]

Using ad hoc simple foot ratios based upon the English foot, in the service of astronomical counting such as: 27 feet representing the lunar month at Crucuno (Carnac) enabling days to be counted in parallel, using Iberian feet; nodal units such as Thom's early megalithic yard of 2.72 feet; and the yard of 3 feet containing 100 shu.si.[§]

Phase Three: A Foot-Based Metrology for Handling Mathematical Functions

Using ratios of the English foot to approximate to irrational and geometric functions: measures are able to map feet to √2 or its reciprocal, to π, or to other measures related to the models in chapter 2.

---

[*]Corresponding to the work of Heath and Heath (2011) and Heath (2014)

[†]See "Reading the Angelic Mind" in chapter 1, p. 14.

[‡]Corresponding to the work of Alexander Thom (1967, 1971, 1978, 1980)

[§]More on the types of megalithic yard and the shu.si can be found in the box above (p. 237) called "Overview of Megalithic Units of Measure."

The English foot was long enough to form fractional ratios in which the number field could be expressed as a calculating tool, since the measurement of a length using a different ratio of foot length gives a result in which the original measurement has been multiplied by the denominator of the fraction and divided by the numerator. Thus, 9 feet becomes 8 feet of the ratio 9/8. The initial approach to such ratio-based feet was to build right triangles using English feet so that the foot of 9/8 feet emerged from a base length of 8 feet and hypotenuse of 9 feet. Above each foot on the base were 8 demarcated feet of 9/8 feet (see fig. 2.1, p. 34), and there are strong reasons to suspect grids of unit squares were in use to form triangles since the right angle is native to such grids, which are also conceptually adapted to studying pure numerical interactions in space.

Phase Four: A Metrology of Foot-Based Modules and Microvariations*

Foot modules evolved as a general-purpose toolkit involving only the prime numbers {2 3 5 7 11}: the systems of root measures using right triangular ratios from the common English foot standard; a common grid of microvariation within each module, applicable to geodetic surveying and modeling; and some less common microvariations such as 225/224 and 81/80.

Phase Five: The Foot-Based Metrology Discovered from the Historical Period†

The historical measures were found through exploration of the geographical regions after which they were named, such as measuring sticks, anthropomorphic sculptures, objects whose size was noted in antiquity, modern-era survey measurements (e.g., by Petrie and Thom), and through inductive metrology, measuring surviving sites and artifacts.

THREE KEY TOPICS

The Astronomical Megalithic Yard

Forming rectangles with a right angle was thought to have been done by using the simplest Pythagorean triangle with sides {3 4 5}, perhaps using a knotted belt

---

*Corresponding to the recent books about ancient metrology from John Michell (1981, 2008) and John Neal (2000, 2016, 2017)

†Corresponding to Petrie (1877) and Berriman (1953)

with 13 knots so that knots 0 and 12 were held together while knots 4 and 7 were held apart to form a right angle to the base of 4 units. However, it has recently become clear that the AMY of 19.008/7 feet, which was derived from the ratio between the solar and lunar years over 19 years, has helpful properties with regard to forming squares of side length 1 MY. The diagonal length is exactly 4 feet of 24/25 feet (historically called the Roman foot).

This meant that when counting months using MYs, a grid of 12 squares was easily constructible using a rod of 1 AMY and another rod of 4 Roman feet, the length of which is √2 larger than the AMY. In fact, two AMY measuring rods could be rotated about one of their their ends, to form a right angle by placing a 4 Roman foot rod between the unconnected ends. A rope could also have three divisions between four knots of length {AMY AMY (4 Roman feet)} or {AMY (4 Roman feet) AMY} to make half a square of side length AMY, in the same way.

This highlights the strange fact that megalithic period astronomical discoveries were often facilitated by shortcuts inherent in the numerical nature of the subject being studied and, furthermore, that the means necessary for their study were possible using Stone Age skills.

The Harmonic Synods of the Outer Planets

Having reached a solar-lunar calendar using the Lunation Triangle or four-square rectangle, the base of such geometries was the lunar year, which also forms harmonic rather than geometrical relationships with the visible outer planets, Jupiter and Saturn. Chapter 5 gives an account of how the loops of these two planets interacted with the full Moon and enabled their synodic periods to be counted using horizon astronomy. The triangles of their synods to the lunar year do not have third sides with integer lengths. Jupiter's 9 to the lunar year's 8 units of 1.5 lunar months lead to a right triangle with just over 4 units as the third, smallest side. But again, as per figure 1.1, the grid technique makes the establishment of the right angle simple so that the third side just exceeds the 4 side of an 8 by 4 rectangle.

Most harmonic ratios are superparticular, in that when the difference (seen as a triangle) is 1 unit, then the two longest sides are integers in the ratio of the interval. This is different from the solar-lunar ratio expressed in the Lunation Triangle, where the 1-foot difference between the longer sides (over

1 year), divided into those sides, gives their normalized ratio as 32.618 feet to 33.618 feet, noting that 32.618 inches is the AMY length in inches. These sides are not then integers. In contrast, the synods of the outer planets form harmonic ratios of 9/8 and 16/15, which are integers, though in different units.

Using AMY squares, one can form a four-square rectangle out of 12 units horizontally and 3 units in the vertical, where each of the four squares are 3 by 3 (that is, 9) AMY squares. The first 8 squares have a base of 8 AMY and a height of 3 AMY. The base therefore equals the 8 units found within the denominator of 9/8. Pythagoras has the third side as $\sqrt{(9^2 - 8^2)} = \sqrt{17}$, while the diagonal length of the four-square triangle is $\sqrt{(4^2 + 1)} = \sqrt{17}$, *the same.* One can therefore take the length of the lowest set of 4 AMY triangles and erect the right length for the 9/8 right triangle above the end of 8 lunar months, that is, 8 AMY (see fig. A1.1). Again, a simple procedure using already established precursors enabled the megalithic astronomers to realize the ratio between the Jupiter synod and the lunar year using the strangely common length of the four-square diagonal, over one-third of a lunar year, and the 9/8 triangle's third side.

The synod of Saturn is 12.8 lunar months long, while the octave defined by Jupiter as a musical fifth and the lunar year as musical fourth defines a tonic

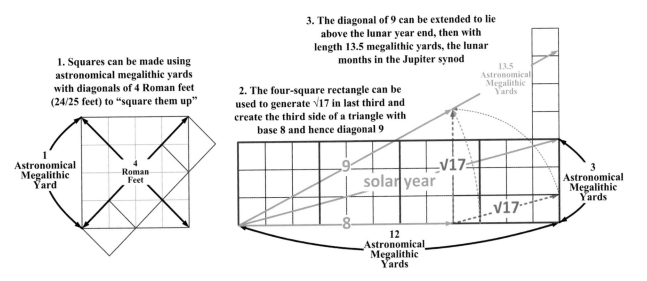

FIGURE A1.1. How the AMY can easily form squares, then grids, based on the four-square rectangle.

range for the octave as 9 to 18 lunar months. The interval of Saturn to the Moon is the tonic's famous opponent, the tritone, which is ideally $\sqrt{2}$, the same ratio as the diagonal of a square to its side length. The square root of two is irrational and cannot be exactly modeled by the integer methods of metrology. Its approximation of 4 Roman feet over a side length of the AMY is, however, almost exact, using two prime number factors {7 11} beyond the {2 3 5} of classic harmony. But Saturn's synod expresses a harmonic approximation to $\sqrt{2}$ as 64/45 = 1.4<u>2</u>, so that, while 9 lunar months × $\sqrt{2}$ = 12.727 lunar months, the Saturn synod was adapted to 12.8/9 (in integers 64/45) = 1/4<u>2</u>. This means one can only observe the defining full Moon in its loop, to measure the synod, every 5 synods. Jupiter, being 13.5 lunar months long, can be measured using the full Moon every 2 synods.

The last section of chapter 1, "Megalithic Use of Angelic Ratios," shows how this phase was able to resolve musical harmony, from which many of the root values of later modules came to express harmonic interval ratios, exactly because these simpler ratios prefigure the more complex metrological ratios employing primes {7 11}. Of particular interest is the use of proximation, a term used for the combined use of two ratios that are approximately the reciprocal of each other. At Crucuno (chapter 1), this was magnificently demonstrated through using 27 feet to represent the lunar month, enabling days within months to be counted as (what became) Iberian feet of 32/35 feet, a mysterious possibility only due to the fact that 32 lunar months equal 945 days (to within half an hour), leading to an approximation to the month of 29.53125 days. The proximation lies between the 29.53059 days of the lunar month, divided by 27 (1.0937), nearly equaling 35/32 (1.09375), these two ratios being unequal by only one part in nearly 45,000!

Ancient World Geodesy

This technique of proximation transformed metrology from being an astronomical tool into an investigation into the size of the Earth. The Equal Perimeter Model of chapter 1 requires the use of primes {7 11} to model π, the circular constant whose irrational value can be modeled using approximations employing primes {2 3 5 7 11}, the standard being 22/7. Two other approximations were introduced, namely 25/8 (3.125) and 63/20 (3.15), in order to create the proximate grid ratios 441/440 (John Neal's *standard* classification) and 176/175 (Neal's *canonical*), which characterized phase four's modular feet. Their sum gives 126/125 (Neal's *standard canonical*), and further application of 176/175 arrives at 1.01376 (or ratio 3168/3125, Neal's *standard geographical*). Each of the ratios is a proximation of two values of π; for example, 176/175 is 22/7 (good π) times 8/25 (not so good π), these two approximations differing by (of course) 1/175th part. Since most monuments analyzed in this book come from phase four, their metrology inherited aspects of phase one, two, and three's metrology, after which, when historical metrology was added (without a knowledge of these previous phases) metrology became quite arcane, with its measures only fleetingly interconnected, the structure unclear, and its assessment of an early development by the megalithic period (as in Thom's work in the earlier phases), unlikely. Recent treatments of ancient metrology (Michell and Neal), while resolving historical metrology (Berriman), found the simpler ratios, between different modules of the early phase, a fact with no history to it. We have instead chosen to see the development of the subject as starting from the use of simple ratios in astronomy. This was further developed, again with remarkable genius, to employ proximation in the service of geodesy. The new metrological model superceded the Equal Perimeter Model described in chapter 2 with one more accurate, and in chapter 3 "Measurements of the Earth," we find a perfect example of the metrological model within Rome's Pantheon.

THE FOOT-BASED MODULES OF ANCIENT METROLOGY

We will now explore geodetic metrology through two modules recently documented in Neal's later work (*Ancient Metrology,* vols. 1–3), in support of my finding these as present in a range of ancient monuments, as described within the main body of this book.

The Byzantine Module

1. The geodetic variations of the English foot are special in generating many unique and opportune ratios. The exemplar is the Byzantine foot* of 128/125, which belongs to the root variation 64/63, as per table A1.1.

---

*Found at the Hagia Sophia at Constantinople, see p. 85, whose dome is 100 Byzantine feet across.

TABLE A1.1
METROLOGICAL MODULE OF
THE BYZANTINE FOOT OF 128/125 (1.024) FEET

CALLED	Byzantine Foot	"Reciprocal"		"Canonical"
"Standard"	× 441/440	—	56/55	128/125
"Root"		100/99	**64/63**	—
	English Foot	× 175/176	**(1 foot)**	× 176/175
"Root"		—	**63/64**	99/100
"Least"	× 440/441	125/128	55/56	—
	Inverse Byzantine Foot	"Reciprocal"		"Canonical"

2. Each such module of variations has its inverse twin module whose root is the reciprocal of the other's root, fractionally "on the other side" of the English foot, in this case, 64/63 has a symmetrical twin module of 63/64, in which the ratios are numerically similar but reversed and, in table A1.1, opposite each other.

3. The two variations of 64/63 and 63/64 are closer to the English foot than the regular modules discussed below because they employ larger integers in their fractions, so they approach the English value of 1. But they seem to have been similarly applied, like the regular module ratios, to be rational to the English foot as the root of the system, the cornerstone of a unified system of metrology.

4. The Byzantine foot (128/125 = 1.024 feet) that derives from this 64/63-foot module is also a musical ratio, called the minor diesis. It therefore could be used to reduce a measurement (or an aggregate measure such as a mile) by 125/128.* This demonstrates a fundamental feature of measures: that applying a measure to a measurement is the equivalent of multiplying the measurement's *numerical value* by the measure's inverse ratio. This perhaps explains the title of Neal's earlier book, *All Done with Mirrors,* as a symmetrical "mirror doctrine" of symmetrical reciprocation.

---

*This minor diesis of three major thirds is what separated the "Feathered Serpent" of the inner planets from life on Earth (see fig. 5.12, p. 104).

5. The ratio 100/99 seems to have been useful at Angkor Wat, as multiplying this ratio by 7/5 (the most basic approximation of the square root of 2 as 1.4) then makes 1.<u>41</u> feet (140/99), just two parts in 39,000 less that the irrational value. As noted in chapter 7's discussion of ad quadratum (see "The Models Within," p. 173), a measure giving the square root of 2 relative to the English foot would render both its squares and nested diamonds rational as found in the AMY and Roman foot relationships of figure A1.1 (p. 243).

The Samian Module

This brings us to another important variation of the English foot, the Samian foot referred to by Herodotus of Samos in his *Histories,* who measured the 756-foot "index" side-length of the Great Pyramid as 800 of his own feet. This gives us 756/800 as 189/200 equaling 0.945 feet, which, varied by 440/441, arrives at the root of another English variation, the Samian foot of 33/35 feet.

The root Samian of 33/35 feet enabled the accurate approximation to the square root of 2 as 99/70 as its cubit, the reciprocal of the square root of 2. This would have led the metrologist to either (a) use it to measure the side of a square, which, in English feet, will then invert it, or (b) seek the inverse Samian measure, *which will directly deliver it.* The historical survival of the ratio 33/35 as this foot implies that the mirror doctrine was employed in secrecy, focusing later minds to the abhorrence of the square root of 2 in early Pythagorean groups, Pythagoras himself coming from Samos.

TABLE A1.2
METROLOGICAL MODULE OF
THE SAMIAN FOOT OF 33/35 (0.<u>9428571</u>) FEET

CALLED	Inverse Samian	*"Reciprocal"*		*"Canonical"*
"Standard"	× 440/441		200/189	
"Root"	× 1		**35/33**	16/15
	English Foot	× 175/176	**(1 foot)**	× 176/175
"Root"	× 1	15/16	**33/35**	
"Standard"	× 441/440		189/200	
CALLED	Samian Foot	*"Reciprocal"*		*"Canonical"*

In the inverse Samian portion of the table, the musical ratio of 16/15, a semitone, is the canonical variant, which means the use of the reciprocal variant of the root Samian of 33/35 can express the semitone by a measurement, made using it, being remeasured in English feet (the mirror doctrine again). It can be no coincidence that Saturn's synod forms a square root relationship to the musical octave of the lunar year's cubit of 18 lunar months, while that synod relates to the lunar year itself as being 16/15 larger than 12 lunar months.

The Samian foot is also important for expressing the relationship of the day length and the lunar month, which is 945 days in 32 months,* or 945/32 = 29.53125 days. The standard variant is 189/200, and this is 0.945 feet. One thousand such feet are 945 feet and 1000/32 are 31.25 Samian feet, which are then 29.53125 feet. So, 31.25 standard Samian feet could then also represent the lunar month. Twelve lunar months would be 375 standard Samian feet, giving (times 0.945) 354.375 day-feet instead of 354.367—better than one part in 44000, again!

The Vedas might have put such a metrological process in this way:

> *Weave back, weave forth.*
> *Man stretches it and man shrinks it;*
> *Even the vault of heaven he has reached with it.*
>
> RIG VEDA 10.130.1

These simple measures only gave such powers because they, in fact, lie behind the phenomenal realities of a complex but ordered celestial world that is governed by rational ratios between its key time cycles.

The Historical Modules of Metrology

It now remains to list the conventional modules of historical metrology and their ancient microvariations, with the caveat that other modules almost certainly existed (such as the Byzantine and Samian) as did other microvariations of them (such as 225/224 and 81/80).

---

*This relationship is explicit within the Le Manio Quadrilateral, where there are 945 inches from the sun gate to stone 32 of the Southern Kerb, which is how I came across it.

TABLE A1.3
METROLOGICAL MODULES KNOWN TO HISTORY*

Assyrian / Mycenean Foot					
0.891824	0.896920	0.902045	0.907200	0.912384	*"Standard"*
0.988669	0.994318	**0.900000**	0.905143	0.910315	*"Root"*
"Least"	*"Reciprocal"*	**9/10**	*"Canonical"*	*"Geographical"*	
Iberian Foot					
0.905980	0.911157	0.916364	0.921600	0.926866	*"Standard"*
0.988669	0.994318	**0.914286**	0.919510	0.924765	*"Root "*
"Least"	*"Reciprocal"*	**32/35**	*"Canonical"*	*"Geographical"*	
Roman Foot					
0.951279	0.956715	0.962182	0.967680	0.973210	*"Standard"*
0.988669	0.994318	**0.960000**	0.965486	0.971003	*"Root "*
"Least"	*"Reciprocal"*	**24/25**	*"Canonical"*	*"Geographical"*	
Common Egyptian Foot					
0.970693	0.976240	0.981818	0.987429	0.993071	*"Standard"*
0.988669	0.994318	**0.979592**	0.985190	0.990819	*"Root"*
"Least"	*"Reciprocal"*	**48/49**	*"Canonical"*	*"Geographical"*	

*1. The tables below have the values increasing from right to left, and from the row below (called "Root") to the row above ("Standard").

2. In John Neal's books, these rows get larger downward but this seemed counterintuitive for new readers.

3. The vertical increase in each foot between columns is 441/440—the "standard" classification of Neal.

4. From left to right within columns, feet increase by 176/175—from "Least" to "Geographical."

5. The root of each module is central to the lower rows and in boldface to mark the "root" value of each module relative to the English foot, which is the "root" value 1 for the Greek module.

6. Not all the foot values have been found to exist within monuments etc., but the modules are well attested historically with one or more instances, in the field.

TABLE A1.3 (cont.)
METROLOGICAL MODULES KNOWN TO HISTORY

Greek/English Foot					
0.990916	0.996578	1.002273	1.008000	1.013760	*"Standard"*
0.988669	0.994318	**1.000000**	1.005714	1.011461	*"Root"*
"Least"	*"Reciprocal"*	**1**	*"Canonical"*	*"Geographical"*	

Common Greek Foot					
1.019227	1.025052	1.030909	1.036800	1.042725	*"Standard"*
0.988669	0.994318	**1.028571**	1.034449	1.040360	*"Root"*
"Least"	*"Reciprocal"*	**36/35**	*"Canonical"*	*"Geographical"*	

Persian/Manx Foot					
1.040461	1.046407	1.052386	1.058400	1.064448	*"Standard"*
0.988669	0.994318	**1.050000**	1.056000	1.062034	*"Root"*
"Least"	*"Reciprocal"*	**21/20**	*"Canonical"*	*"Geographical"*	

Belgic/Drusian Foot					
1.061695	1.067762	1.073864	1.080000	1.086171	*"Standard"*
0.988669	0.994318	**1.071429**	1.077551	1.083708	*"Root"*
"Least"	*"Reciprocal"*	**15/14**	*"Canonical"*	*"Geographical"*	

Sumerian Foot					
1.080999	1.087176	1.093388	1.099636	1.105920	*"Standard"*
0.988669	0.994318	**1.090909**	1.097143	1.103412	*"Root"*
"Least"	*"Reciprocal"*	**12/11**	*"Canonical"*	*"Geographical"*	

Saxon Foot					
1.090007	1.096236	1.102500	1.108800	1.115136	*"Standard"*
0.988669	0.994318	1.100000	1.106286	1.112607	*"Root"*
"Least"	*"Reciprocal"*	**11/10**	*"Canonical"*	*"Geographical"*	

TABLE A1.3 (cont.)
METROLOGICAL MODULES KNOWN TO HISTORY

Yard + Hand Foot

1.101017	1.107309	1.113636	1.120000	1.126400	**"Standard"**
0.988669	0.994318	1.111111	1.117460	1.123846	**"Root"**
"Least"	**"Reciprocal"**	**10/9**	**"Canonical"**	**"Geographical"**	

Royal Egyptian Foot

1.132475	1.138946	1.145455	1.152000	1.158583	**"Standard"**
0.988669	0.994318	1.142857	1.149388	1.155956	**"Root"**
"Least"	**"Reciprocal"**	**8/7**	**"Canonical"**	**"Geographical"**	

Russian Foot

1.156068	1.162674	1.169318	1.176000	1.182720	**"Standard"**
0.988669	0.994318	1.166667	1.173333	1.180038	**"Root"**
"Least"	**"Reciprocal"**	**7/6**	**"Canonical"**	**"Geographical"**	

APPENDIX 2

The Significance of the Moon

A MOON THAT MADE LIFE POSSIBLE?*

Over 4.5 billion years ago, the inner solar system was a jumble of would-be planets and planetoids. It is thought that Earth shared its orbital zone with at least one competitor, about the size of Mars, similarly composed of a heavy metal core and an outer mantle. Both would have been mopping up smaller bodies, but eventually the two collided with each other.

In a vast explosion, the Earth was severely damaged while the energy released caused vast amounts of both planets' surface rocks to be vaporized or projected whole into space. This caused a ring to form around the Earth that quite rapidly accreted (consolidated) into a single body, which soon cooled to form a moon orbiting every 20 days, a mere 2700 kilometers above the Earth's surface. The planet that struck Earth has been called Thea after the goddess that gave birth to the Moon, the latter being called Selene in Greek myth. Meanwhile, the metallic core of Thea was not absorbed by the Earth's core, but instead, significant metal deposits were embedded in the surface layers, a fact that gave the Earth a rich "wedding ring" of workable ore deposits, which was significant to the later metal-working ages.

Such a massive satellite traveling over the Earth caused the whole surface to gravitationally deform below the Moon, but the Earth was then rotating

---

*This section is extracted from my book *Precessional Time and the Evolution of Consciousness*, 64–67.

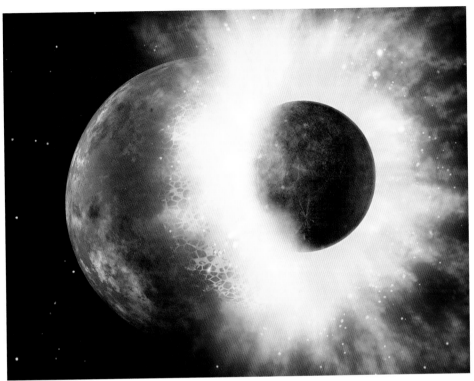

FIGURE A2.1. The collision of Earth and Thea.

every six hours so that this bulge would always be ahead of the Moon. Just as with tides today, but then much more strongly, the Earth's rotation transferred energy to the Moon, causing it to accelerate and take an ever-higher orbit. However, this was not before the Moon had kneaded all the surface rocks. This type of lunar influence then continued in an unusual way.

Around 4 billion years ago, the orbits of Jupiter and Saturn aligned so as to create a slingshot for solar system bodies that had not yet been incorporated into planets. This "Late Great Bombardment" proceeded to strike the Moon rather than the Earth, and this protective role is thought to have saved the Earth from damage to its nascent resources, such as the water present on its surface. The recognizable face of the Moon was largely created at this time, as craters and "seas" of molten basalt formed from this bombardment.

The Moon was further accelerated to an orbital distance of 320,000 kilometers by 3 billion years ago, and this meant that its tidal effects were no longer strong on the mantle rocks, but instead, the seas of that period experienced massive tides, hundreds of meters high. These must have been like continuous tsunamis racing around the globe. Meanwhile, due to the still great rotational speed of the Earth, these tides occurred many times a day, and also, the early atmosphere was whipped up by the Coriolis effect so as to create continuous hurricane-speed winds. The extreme ocean tides caused massive erosion and mineralization of the seas, forming a massive number of chemical scenarios that could even have been responsible for the creation of life in the form of primitive replicating molecules.

Even today, volcanoes and earthquakes are thought to trigger eruptions and the release of seismic energy built up in the Earth's unique tectonic plates. These plates themselves could be an artifact, in part, of the Moon's kneading of the Earth, and we can see that on Mars, where any plate activity ceased billions of years ago as the mantle became stuck to a solidified core, probably through Mars' lack of a large moon.

While the original collision almost certainly caused the high spin of the Earth, it also created the tilt of the axis on which the Earth rotates. This tilt set up the seasonal conditions on Earth, which are so important for life's varied habitats. However, this tilt would not have been stable without the large Moon that also resulted. Our large Moon stabilizes the tilt by shielding the Earth from the small chaotic forces the Earth experiences due to the other planets. Mars is particularly vulnerable, and its tilt has varied over millions of years by about 30 degrees. That is, the Moon, by adding a large systematic component to the precessional forces, prevents planetary chaotic resonances from affecting the Earth.

The effect of the seasons, which are therefore maintained by the Moon, is joined by the extra tidal effect the Moon has on the seas and oceans. These tides create an extensive area of important habitat within the tidal ranges found on our coastlines. These are very biodiverse and also have led to evolutionary changes as significant as the adaptation of marine animals into land animals.

In summary, therefore, life on Earth would not have been possible without the Moon and the very special itinerary of its genesis and the gradual arrival at the conditions we find today. It all seems a little too special, and this has led to the general recognition that life such as is found on Earth could not have evolved without such a special collision occurring at exactly the distance from the Sun capable of supporting such life.

APPENDIX 3

The Likely Origins of Gurdjieff's Harmonic System

THE ENNEAGRAM WAS NOT AN OCTAVE but, as stated in *Tales,* it was a synthesis of the laws of seven and three. Being that it was a synthesis, Gurdjieff, in 1917, wanted to introduce his students to ancient tuning theory and the structures that develop, through number, within Eternity as deriving from Worlds 3 and 6 within the context of the Universal Will and the dimension of Hyparxis. His cosmology was not like that of Pythagoras but used harmonic numbers in a similar way—as a cosmological system—though its origins lay farther East. The origins of Gurdjieff's ninefold law will probably remain enigmatic, but one can benefit from J. G. Bennett's research into the matter and his collation of generally available facts to resolve historical likelihoods.

A Benedictine monk, Guido of Arrezo (c. 1000 CE), is credited with innovating the hexachordal solfège {*do re mi fa sol la*} during the Middle Ages, yet it was the work of Islamic philosopher Al-Kindi* in ninth-century Babylon, an early Islamic setting with a "Hall of Wisdom" that he presided over, who borrowed many ideas from the Classical Greeks and added many more. Ironically, Islam would soon dispense with such philosophers and their musical theories, but Guido had access to Al-Kindi's Arabic texts via the Jews of Spain, who were ideal translators of Arabic, which, like Hebrew, was a triliteral language.

---

*Al-Kindi was the first great theoretician of music in the Arab-Islamic world. He is known to have written fifteen treatises on music theory, but only five have survived. He added a fifth string to the oud. His works included discussions on the therapeutic value of music and what he regarded as "cosmological connections" of music. Wikipedia, "Al-Kindi."

The Enneagram was probably labeled with the solfège system (and its curious starting of new octaves at points 3 and 6) when Gurdjieff first encountered it. This might have been in Bokhara, Uzbekistan, the center of the Naqshbandi Order.* The parallel between cosmic octaves, solfège's implied diatonic scale, and the Enneagram was perhaps what is meant by Al-Kindi's "cosmological connections" of music, or more likely, the latter could be the Pythagorean harmony of the spheres.

Later, Bennett found a possible source of Gurdjieff's science of vibrations in the school of Ahmad Yasawi (d. 1169). He says:

> Ahmad Yasawi's central school in Tashkent . . . is of special interest to followers of Gurdjieff's ideas because it was the main repository of the science of vibrations expressed partly through dance and music and partly through the sacred ritual that came from the Magi. This science distinguished the Yasawis from the main tradition of the Masters.[1]

West of the Caucasus, the Pythagorean number sciences had the ancient tuning theory, but in central Asia, an ancient "science of vibrations" understood how energies could be transformed within three cosmic octaves between the Sun and Moon.

> In this way, we reach the conclusion that the knowledge that Gurdjieff afterwards taught as his "Ideas" came from putting together two halves of a single truth. One half is found in the Western, chiefly Platonic, tradition and the other half is in the Eastern, chiefly Naqshbandi [and Yasawi] tradition. This fusion of two halves is strongly hinted by Gurdjieff in the story of the Boolmarshano in Chapter 44 of *Beelzebub's Tales*.[2]

It therefore seems likely that Gurdjieff had inherited aspects of Eastern alchemical ideas about vibrations but that his students would need some background in ancient tuning theory to understand his Worlds and the Enneagram. He seemed to drop this approach in *Tales,* since the diagrams either disappear

---

*The order's name alludes to a seal, pattern, symbolism, (*naqsh*) and a name given to their founder, Bahauddin (1318–1389), whom Bennett equates with the Bokharian dervish Bogga-Eddin of *Tales. Gurdjieff: A Very Great Enigma,* 38.

or are sublimated into textual equivalents, perhaps "burying the dog deeper." There was the danger of mistaking his ideas as Pythagorean, and he stated that his new approach was shaped by the need to understand them rather than know them,* and as Bennett stated, understanding, unlike knowledge, requires a structure of will.

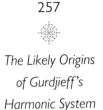
> We are . . . in a stage of confusion due to the inadequacy of our modes of thought. We continue to think in terms of atomic concepts linked by logical implications and empirical laws. This approach can never lead to the understanding of structures whose significance lies in their organized complexity, not in their susceptibility to destructive analysis into elements and laws. We have seen in the earlier chapters, that understanding is the subjective aspect of will and knowledge is the subjective aspect of **Function**.† We can "know" structures only in their functional properties; whereas we "understand" them in their working. This working is very much more than actualization in time, for it concerns what things **are** and not simply how they **change**.[3]

The groups in St. Petersburg and Moscow did not have any clue as to numerical tuning theory, despite some having been educated in musical forms and playing instruments within the equal-tempered world of major diatonic scales within twelve keys. On page 126 of *In Search*, Gurdjieff used the inappropriate numbers from 1000 to 2000 to explain how octaves and scales worked.‡ These numbers lack the prime number 3 and so are alien to numerical harmony, where an octave's *limiting number* (high *do*) is crucial to the formation of tones within its octave.

Octave doubling obviously manifests the prime number 2, the first true *interval* in which doubling creates a region that can only be entered through *larger* prime numbers that can divide up the region of octave doubling. The

---

*"We can 'know' structures only in their functional properties; whereas we can 'understand' them in their working . . . [in] what things **are**."

†Knowledge was defined as the ordering of **Function**. Ordering is an operation performed on the data, whereas understanding is a transformation within the data. Bennett, *Dramatic Universe*, vol. 4, 8; vol. 1, 62–64.

‡Gurdjieff's number may well have been changed by the compilers of *In Search of the Miraculous* (on whose excellent work we depend) because they did not know about harmonic numbers requiring primes {3 5} in their factorization.

prime numbers 3 and 5 "get into" an octave by *dividing it with smaller ratios between 1 and 2*. But the limiting number of 2000 has no prime factors of 3 to "give" to any new integer tones within its octave range from 1000 to 2000, so whole number tones of the Pythagorean kind would not be possible using integers. One concludes from this that the poor number of 2000, presented as the limit, was either (a) not correctly remembered by the students (which is unlikely) or (b) was *deliberately inadequate* for scale formation so that only the diligent would calculate the correct octave range. Gurdjieff says,

> The differences in the notes or the differences in the pitch of the notes are called intervals. We see that there are three kinds of intervals in the octave: 9/8, 10/9, and 16/15, which in whole numbers correspond to 405, 400, and 384.[4]

There is only one number that can form these three intervals to these three numbers: 360, and 360 is low *do* (in tuning theory) for the lowest possible limit forming five different scales, namely the high *do* of 720. This limit was used in the Bible's earliest chapter, Genesis (written c. 600 BCE in Babylon), to define Adam (whose Hebrew letter-numbers sum to 45* which can then double four times to 720).

The mountain for 720 (see fig. A3.1) shows the initiation of three scales from D (= 360) to E as 405 (Mixolydian scale), to e as 400 (Ionian) and e♭ as

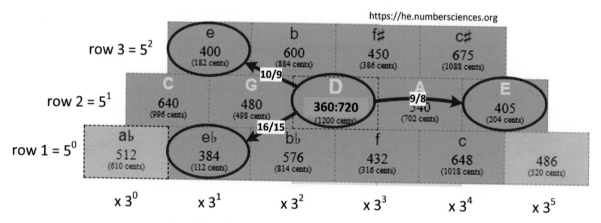

FIGURE A3.1. The harmonic mountain and the tone circle of 720.

---

*A.D.M = 1.4.40 = 1 + 4 + 40 = 45, or in position notation 1440, which is 32 × 45.

384 (Phrygian scale).* This is a very significant situation within numerical tuning theory, for only in the octave from 360 to 720 can the only three intervals out of which scales are made be resolved in this way from a single starting number of 360. This is proof positive that Gurdjieff understood ancient numerical tuning theory.

However, this was a sideshow for what was being explained as Worlds whose limits contain only one prime number 3 rather than the two: $3 \times 3 = 9$, times 5 to make 45, the number of Adam. Gurdjieff's scheme of Worlds 12, 24, 48, and 96 was, he said, a series of octaves naturally present in the early numbers below 96, which are between Worlds that in the harmonic model for 72:1440 have numbers associated with the lunar year, that is, 960.

Hence, the numbers of Gurdjieff's Worlds did not create octaves *by the methods of limits* but rather through the properties of the first 96 numbers (see chapter 9, "The Enneagram of Life"). Gurdjieff labeled the Ray of Creation with solfège symbols, yet the key was also shown in the Enneagram, in which every world or starting point is like a *do*. Above the Sun, the Ray of Creation loses its coherence since Worlds 3 and 6 are the three forces and their permutations to form six kinds of dynamism. At Worlds 12 to 24, music starts to emerge, and in Worlds 24 to 48, the major scale emerges as the series {24 27 30 32 36 40 45 48}, which is {*do re mi fa sol la si do*}, as is well known to tuning theory as the earliest number set able to form that scale. Every doubling thereafter is certain to double all the note numbers, making C-major available as, for example, {48 54 60 64 72 80 90 96}. In the Enneagram, the shocks appear to occur (at point 3) between the interval 72 to 80 and (at point 6) between the same interval, 80 to 72.

In the Russian lectures there were hints of the law of ninefoldness:

In the study of the law of octaves it must be remembered that octaves in their relation to each other are divided into fundamental and subordinate. The fundamental octave can be likened to the trunk of a tree giving off branches of lateral octaves. The seven fundamental notes of the octave and the two

---

*It is common practice in music theory to use lower case for note classes not belonging to the Pythagorean scale but, instead, differ by the synodic comma introduced by the scales of Just intonation 81/80, which employ prime number {5} to mitigate the growth of powers of prime number {3}, since 81 is the fourth power of 3 while 80 contains a single 5 ($\times$16).

"intervals," the **bearers of new directions**, give altogether nine links of a chain, three groups of three links each. (Emphasis added.)[5]

These *bearers of new directions* turn the law of seven into the law of nine, a law that can double further, at least in the sublunary world, by further doubling according to the "law of vibrations," as is seen in essence classes below the invertebrates (see fig. 9.7). Bennett more broadly called that the "germinal essence," therefore including seeds. The major scale and its inverse, the Phrygian, reign supreme by dint of their ability to arise from the doubling of the number 3, as {3 6 12 24 48 96 192 384 768 . . .}, as a structure of will.

In the world of harmony and Pythagorean scales, the heptachord was a different scale (called Dorian) that was generated by tuning successive fifths, as did the Chinese. Dorian is naturally played on a keyboard using D, which is the white key at the center of symmetry between the other white and black keys (fig. 7.24). This symmetry of the Pythagorean heptachord is broken by starting at C, from which the major scale is formed, uniquely and solely, from the white keys. This accident, corresponding to the cosmic scales, came about through church music, which preferred C major leading to the formation of modern musical notation and the keyboard. The major scale further developed when equal temperament emerged, breaking with the numerical laws of tonality and seeking instead a compromise that enabled the modulation of keys based on the major scale and thereby the creation of modern music in which the large orchestra became possible.

Therefore, Gurdjieff would have had to tread carefully in his use of a musical theory to explain his cosmology, and, in the end, he found other ways to express this without comparing his cosmology of will to the musical manifestations of Being. His talks were also in danger of being drawn into ancient tuning theory which, latterly, has been found to have a strong but different form of esotericism, which certainly influenced the Abrahamic religions. As already stated, religion is a path of Being and not Will: it is not that saints and yogis did not express will, it is simply that the cosmology on which their efforts focused was based on Being rather than Will. Gurdjieff's musical cosmology was something new.

Notes

CHAPTER 1. RATIOS OF THE ANGELIC MIND

1. Fred Hoyle, "The Universe: Past and Present Reflections," 8–12.
2. Heath and Heath, "Origins of Megalithic Astronomy."
3. Heath, *Sacred Number and the Lords of Time,* 87 onward.
4. First presented in Neal, *All Done with Mirrors.*
5. Neal, *Ancient Metrology,* 2:11.

CHAPTER 2. MODELS OF TIME AND SPACE

1. Robin Heath, *Sun, Moon and Stonehenge.*
2. Heath and Heath, "Origins of Megalithic Astronomy."
3. For more on the subject of Carnac's encounter with time, space, and geometry, see Heath, *Sacred Number and the Lords of Time,* chap. 2.
4. Michell, *Jerusalem.*
5. See Heath, *Sacred Number and the Origins of Civilization,* 59 and fig. 4.3.
6. Crowhurst, *Carnac.*

CHAPTER 3.
MEASUREMENTS OF THE EARTH

1. Allen, *Astrophysical Constants,* 148.
2. A fuller treatment is found in Michell, *Dimensions of Paradise;* on the factorial model, see Michell, *Sacred Center,* chap. 7, and Heath, *Sacred Number and the Lords of Time,* chap. 8.

3. Berriman, *Historical Metrology.*

4. Michell, *Ancient Metrology,* 47.

5. Michell, *Ancient Metrology.*

6. Michell, *Dimensions of Paradise,* see p. 44 and index of numbers for 3168.

CHAPTER 4.
TEMPLES OF THE EARTH

1. Wikipedia, "Sublunary Sphere," which references Gillespie, *The Edge of Objectivity,* 14.

2. Mayer-Baer, Kahi, *Music of the Spheres and the Dance of Death.*

3. Wikipedia, "Flinders Petrie," accessed April 23, 2020.

4. Kappraff and McClain, "Proportional System of the Parthenon."

5. In the ancient model of the Earth, three different approximations of π were used to allow for the distortion of the rotating planet over what its mean, or perfectly spherical, circumference would be. This information was first published as a pamphlet by John Michell, *Ancient Metrology,* 1981, then as a book *Dimensions of Paradise,* 2008, and in its full and final form by John Neal *All Done with Mirrors,* 2000; and *Ancient Metrology,* three volumes: 2016, 2017, and 2020 (forthcoming).

6. Dhavalikar, *Sanchi.*

PART TWO. THE COSMIC INDIVIDUALITY

1. Rogers, "Origins of the Ancient Constellations."

CHAPTER 5.
HARMONY OF THE LOCAL COSMOS

1. Ancient Near Eastern musicologist Richard Dumbrill, private communication. See also Dumbrill, "The Truth about Babylonian Music."

2. Levy, *Theory of Harmony;* and Levarie and Levy, *Musical Morphology.*

3. Dumbrill, "Four Tables."

4. Robin Heath, *Sun, Moon and Stonehenge,* 102–4.

5. Thom, Thom, and Burl, *Megalithic Rings,* 47.

6. Hurwit, *Art and Culture of Ancient Greece,* 74–75.

7. Hurwit, *Art and Culture of Ancient Greece,* 76.

8. Hurwit, *Art and Culture of Ancient Greece,* 76–77.

9. For a general background, see Wikipedia, "Heraion of Samos."

10. Heath, *Harmonic Origins of the World,* 181, last paragraph.

11. Sugiyama, "Teotihuacan City Layout," especially fig. 11.9.

CHAPTER 6. THE SAVIOR GOD

1. Bennett, *Dramatic Universe,* 4:259–84.
2. Bennett, *Dramatic Universe,* 4:334–38.
3. See systematics.org and Bennett, *Dramatic Universe,* vol. 3, chap. 37, "The Structure of the World."
4. Allen, "Romano-British Periods."
5. de Santillana and von Dechend, *Hamlet's Mill.*

CHAPTER 7. PAVEMENTS OF THE SAVIOR

1. Foster, *Patterns of Thought.*
2. Geoffrey of Monmouth, *History of the Kings of England,* 1127.
3. Taylor, *How to Read a Church.*
4. Such as Michell, *The View over Atlantis,* an early-1970s bestseller, reprinted in 1983 as *Glastonbury: The New View over Atlantis.*
5. See Roberts, *Glastonbury: Ancient Avalon, New Jerusalem,* 176–77, "Addendum: Sources of the New Jerusalem Diagram."
6. Arthur L. Howe, *The Painting and the Floor—The Glastonbury Code,* (film) 2007.
7. Michell, *Glastonbury: The New View over Atlantic,* chap. 6.
8. Foster, *Patterns of Thought,* is a good treatment on the Westminster pavement. The classic paper is Wander, "Westminster Abbey Sanctuary Pavement."
9. Dudley, *Canterbury Cathedral.*

CHAPTER 8. THE FOCAL BUILDINGS OF ISLAM

1. Heath, *Harmonic Origins of the World,* fig. 1.5 and the section starting on p. 20.
2. A summary of traditional stories abstracted from the introduction of Esin, *Mecca the Blessed.*
3. McClain. *Meditations through the Quran.*
4. McClain, *Meditations through the Quran,* 78–79.
5. McClain, *Myth of Invariance,* 126.
6. Wikipedia, "Monolith (Space Odyssey)."

CHAPTER 9. COSMOLOGICAL NUMERACY

1. Placed into the mouth of Kurdish philosopher in *Beelzebub's Tales,* 1094.
2. Gurdjieff, *Beelzebub's Tales,* 88 and 1222.

3. Bennett, *Dramatic Universe,* vol. 1, glossary, p. 522, and section 1.3.2.

4. Bennett, preface to *Dramatic Universe,* vol. 4.

5. From an early version of Bennett, *Dramatic Universe,* vol. 3, reprinted as *Creation.*

6. Bennett, *Creation,* 97.

7. Bennett, *Creation,* 106–9.

8. From an early version of Bennett, *Dramatic Universe,* vol. 3, reprinted as *Creation.*

9. Ouspensky, *In Search of the Miraculous,* 322.

10. "Pentad," systematics.org.

11. Bennett's full exposition of pentads within the diagram can be read in *Gurdjieff: Making a New World,* app. 2.

12. Bennett, *Gurdjieff: Making a New World,* app. 2.

13. *Dramatic Universe,* 4:307–10, 324, 438.

14. Gurdjieff, *Beelzebub's Tales,* 753.

15. Gurdjieff, *Life Is Real Only Then,* 24.

16. Gurdjieff, *Beelzebub's Tales,* 753.

17. "Ennead," Systematics website, systematics.org.

18. Gurdjieff, *Beelzebub's Tales,* 771.

19. Douglas, *Thinking in Circles.* 2010.

20. Bennett, *Creation,* 109–10.

POSTSCRIPT. EMERGENCE

1. John 3:3, King James Version.

APPENDIX 3.
THE LIKELY ORIGINS OF GURDJIEFF'S HARMONIC SYSTEM

1. Bennett, *Masters of Wisdom,* 131.

2. Bennett, *Gurdjieff: A Very Great Enigma,* 41.

3. Bennett, *Dramatic Universe,* 3:8.

4. Ouspensky, *In Search of the Miraculous,* 126.

5. Ouspensky, *In Search of the Miraculous,* 134.

Bibliography

Allen, C. W. *Astrophysical Constants.* London: Athlone, 1983.

Allen, J. Romilly. *The Romano-British Periods and Celtic Monuments.* Reprint, Lampeter, Wales: Llanerch, 1992.

Bennett, John G. *Creation.* Compiled by Anthony Blake. Gloucestershire: Coombe Springs Press, 1978. Reprint, Santa Fe, N.Mex.: Bennett Books 2016.

———. *Deeper Man.* Compiled by Anthony Blake. London: Turnstone, 1978.

———. *The Dramatic Universe.* 4 vols. London: Hodder & Stoughton, 1956–1965. Vol. 1: *The Foundations of Natural Philosophy* (1956). Vol. 2: *The Foundations of Moral Philosophy* (1961). Vol. 3: *Man and His Nature* (1965). Vol. 4: *History* (1965).

———. *Enneagram Studies.* Rev. ed. New York: Samuel Weiser, 1983.

———. *Gurdjieff: A Very Great Enigma.* London: Coombe Springs Press, 1963.

———. *Gurdjieff: Making a New World.* London: Turnstone, 1973.

———. *The Masters of Wisdom.* London: Turnstone, 1977.

———. *Talks on Beelzebub's Tales.* Gloucestershire: Coombe Springs Press, 1977. Revised edition, Santa Fe, N. Mex.: Bennett Books, 2007.

Berriman, A. E. *Historical Metrology.* London: J. M. Dent and Sons, 1953.

Blake, Anthony G. E. *An Index to "In Search of the Miraculous."* DuVersity Publications, 1982; available at DuVersity.org.

———. *The Intelligent Enneagram.* New York: Shambhala, 1996.

Crowhurst, Howard. *Carnac: The Alignments.* Plouharnel, France: Epistemea, 2011.

Dhavalikar, M. K. *Sanchi: Monumental Legacy.* Oxford: Oxford University Press, 2005.

Douglas, Mary. *Thinking in Circles: An Essay on Ring Composition.* New Haven, Conn.: Yale University Press, 2010.

Dudley, Colin Joseph. *Canterbury Cathedral: Aspects of Its Sacramental Geometry.* Bloomington, Ind.: Xlibris Corporation, 2010.

Dumbrill, Richard J. "Four Tables from the Temple Library of Nippur: A Source for 'Plato's Number' in Relation to the Quantification of Babylonian Tone Numbers."

International Conference of Near Eastern Archaeomusicology, *Archaeomusicological Review of the Ancient Near East* 1:27–38 (2008); pdf available at academia.edu.

———. "Music Theorism in the Ancient World." In Proceedings of the International Conference of Near Eastern Archaeomusicology: Université de la Sorbonne, Paris IV, November 25, 26, and 27, 2009, and at the University of London, Senate House, December 13, 14, and 15, 2010, 107–32.

———. "The Truth about Babylonian Music." *Near Eastern Musicology Online* (2017): 91–121; available on academia.edu.

Esin, Emil. *Mecca the Blessed, Madinah the Radiant.* London: Elek Books, 1963.

Foster, Richard. *Patterns of Thought: The Hidden Meaning of the Great Pavement of Westminster Abbey.* London: Jonathan Cape, 1991.

Gillespie, C. C. *The Edge of Objectivity.* Princeton, N.J.: Princeton University Press, 1960.

Gurdjieff, George I. *Beelzebub's Tales to His Grandson.* London: Routledge & Kegan Paul, 1950.

———. *Guide and Index to G. I. Gurdjieff's "All and Everything: Beelzebub's Tales to His Grandson."* Toronto, Canada: Traditional Studies Press, 1971.

———. *Life Is Real Only Then, When "I Am."* London: Routledge & Kegan Paul, 1981.

———. *Meetings with Remarkable Men.* London: Routledge & Kegan Paul, 1963.

Heath, Richard. *The Harmonic Origins of the World.* Rochester, Vt.: Inner Traditions, 2018.

———. *Matrix of Creation.* Rochester, Vt.: Inner Traditions, 2002.

———. *Precessional Time and the Evolution of Consciousness.* Rochester, Vt.: Inner Traditions, 2011.

———. *Sacred Number and the Lords of Time.* Rochester, Vt.: Inner Traditions, 2014.

———. *Sacred Number and the Origins of Civilization.* Rochester, Vt.: Inner Traditions, 2007.

Heath, Richard, and Robin Heath. "The Origins of Megalithic Astronomy as Found at Le Manio: Based on a Theodolite Survey of Le Manio, Carnac, Brittany, 22nd to 25th March 2010." Published in 2011 and available at the Richard Heath section of the Independent Academia website.

Heath, Robin. *Sun, Moon, and Stonehenge.* Cardigan, Wales: Bluestone Press, 1998.

Heath, Robin, and John Michell. *Lost Science of Measuring the Earth: Discovering the Sacred Geometry of the Ancients.* Kempton, Ill.: Adventures Unlimited Press, 2006. Reprint edition of *The Measure of Albion.*

———. *The Measure of Albion.* Cardigan, Wales: Bluestone Press, 2004.

Hoyle, Fred. *On Stonehenge.* San Francisco, Calif.: W. H. Freeman, 1977.

———. "The Universe: Past and Present Reflections." *Engineering and Science* (November, 1981): 8–12.

Hurwit, Jeffrey. *The Art and Culture of Ancient Greece.* Ithaca, N.Y.: Cornell University Press, 1989.

Kappraff, Jay, and Ernest McClain. "The Proportional System of the Parthenon and Its Connections with Vedic India," In *Music and Deep Memory: Speculations In Ancient Mathematics, Tuning, And Tradition: In Memoriam Ernest G. McClain, 1918–2014,* edited by Bryan Carr and Richard Dunbrill, 81–95. London: Iconea Publications, 2018.

Landay, Jerry M. *Dome of the Rock.* New York: Newsweek, 1972.

Levarie, Siegmund, and Ernst Levy. *Musical Morphology: A Discourse and a Dictionary.* Kent, Ohio: Kent State University Press, 1980.

Levy, Ernst. *A Theory of Harmony.* SUNY Series in Cultural Perspectives. Edited by Sigmund Levarie. Albany: State University of New York Press, 1985.

Mainstone, Rowland J. *Hagia Sophia: Architecture, Structure, and Liturgy of Justinian's Great Church.* London: Thames & Hudson, 1988.

Marshack, Alexander. *The Roots of Civilisation: Cognitive Beginnings of Man's First Art Symbol and Notation.* New York: Weidenfeld & Nicolson, 1972.

McClain, Ernest. *Meditations through the Quran.* York Beach, Maine: Nicolas-Hays, 1981.

———. *The Myth of Invariance.* New York: Shambhala, 1976.

———. *The Pythagorean Plato.* New York: Shambhala, 1978.

Michell, John. *Ancient Metrology.* Bristol, England: Pentacle Press, 1981.

———. *At the Centre of the World.* London: Thames & Hudson, 1994.

———. *Dimensions of Paradise.* Rochester, Vt.: Inner Traditions, 2008.

———. *Glastonbury: The New View over Atlantis.* London: Thames & Hudson, 1983.

———. *Jerusalem.* Glastonbury, England: Gothic Image, 2000.

———. *New Light on the Ancient Mystery of Glastonbury.* Glastonbury, England: Gothic Image, 1990.

———. *Sacred Center: The Ancient Art of Locating Sanctuaries.* Rochester, Vt.: Inner Traditions, 2009. Reprint edition of *At the Centre of the World.*

Mayer-Baer, Kathi. *Music of the Spheres and the Dance of Death: Studies in Musical Iconology.* Princeton, N.J.: Princeton University Press, 1970.

Neal, John. *All Done with Mirrors.* London: Secret Academy, 2000.

———. *Ancient Metrology.* Vol. 1, *A Numerical Code—Metrological Continuity in Neolithic, Bronze, and Iron Age Europe.* Glastonbury, England: Squeeze, 2016.

———. *Ancient Metrology.* Vol. 2, *The Geographic Correlation—Arabian, Egyptian, and Chinese Metrology.* Glastonbury, England: Squeeze, 2017.

Ouspensky, P. D. *In Search of the Miraculous.* London: Routledge & Kegan Paul, 1950.

Petri, W. M. Flinders. *Inductive Metrology.* 1877. Reprint, Cambridge: Cambridge University Press, 2013.

Roberts, Anthony, ed. *Glastonbury: Ancient Avalon, New Jerusalem.* London: Rider Books, 1978.

Rogers, John H. "Origins of the Ancient Constellations: I. The Mesopotamian Traditions & II. The Mediterranean Traditions." *Journal of the British Astronomical Association* 108, no. 1 (1998): 9–28 and no. 2 (1998): 79–89.

Sachs, Curt. *The Rise of Music in the Ancient World: East and West.* New York: Norton, 1943.

de Santillana, Giorgio, and Hertha von Dechend. *Hamlet's Mill: An Essay on Myth and the Frame of Time.* Boston: Godine, 1977.

Schultz, Joachim. *Movement and Rhythms of the Stars: A Guide to Naked-Eye Observation of Sun, Moon and Planets.* Edinburgh: Floris Books, 1986.

Smyth, Piazzi. *The Great Pyramid.* 4th ed. New York: Outlet-Random House 1990.

Soothill, William Edward. *The Hall of Light.* Cambridge: James Clarke & Co. Ltd., 1951.

Sugiyama, Saburo. "Teotihuacan City Layout as a Cosmogram." In *The Archaeology of Measurement,* edited by Iain Morley and Colin Renfrew, 130–49. Cambridge: Cambridge University Press, 2010.

Taylor, Richard. *How to Read a Church.* London: Rider Books, 2003.

Thom, A. *Megalithic Lunar Observatories.* Oxford, England: Clarendon 1971.

———. *Megalithic Sites in Britain.* Oxford: Clarendon 1967.

Thom, A., and A. S. Thom. *Megalithic Remains in Britain and Brittany.* Oxford: Clarendon 1978.

Thom, A., A. S. Thom, and A. Burl. "Megalithic Rings: Plans and Data for 229 Monuments in Britain." Vol. 81, *British Archaeological Reports.* Oxford, England: British Archaeological Reports, 1980.

Tompkins, Peter. *Secrets of the Great Pyramid.* Appendix by Livio Catullo Stecchini. New York: Harper & Row, 1971.

Twohig, Elizabeth Shee. *The Megalithic Art of Western Europe.* Oxford, England: Clarendon Press, 1981.

Vincent, Fiona. "A Major 'Lunar Standstill.'" The SAO/NASA Astrophysics Data System website.

Walter, H. *Das Heraion von Samos.* Munich: Piper, 1976.

Wander, Steven H. "The Westminster Abbey Sanctuary Pavement." *Traditio* 34 (1978): 137–56.

Index